Ark of the Covenant

Ark of the Covenant

Landing Dove

12th Book of Life

Nebulous Henctis

Ark of the Covenant, Landing Dove
Written by Nebulous Henctis
Publisher: Living Tree Services
Published in 2025
ISBN# 978-1-965502-08-2

To the Angel of the Church of Corinth
Adamant

Thank GOD for revealing the keys and sharing how to bring them together.

Thanks to family, friends, and everyone else who helped gather and assemble the keys.

Table of Contents

Introduction 1
   What is the Covenant Ark? 1

Chapter 1 5
   Source Frequency of Creation 5
      Locating the Ark of the Testimony 5
      Universal Comprehension 7

Chapter 2 22
   Integrating the Ark 22
      Scriptural Merging Procedures 27
      Consulting the Urim and Thummim 32

Chapter 3 35
   Matrix of the Mother's Womb 35
      Quantum Atomic Grid 37
      Why is Pi 40
      Light Travel 42
      Multiversal Teleportation 45
      Eternal Life as Seen Through Science 48
      Celestial Resonance 50
      Divine Doctrinal Networking 51
      Highway of Holiness 54

Chapter 4 57
   Scientific Proof of Biblical Authenticity 57
      Jacob's Children – Dust of the Earth 57
      Timeline of Evolution 63
      Seven Days of Religious Creation 68

Chapter 5 77
   The Oracle 77
      Walking on Water 77
      Translating the Oracle 82
      Reading the Inner Veil 84
      Keys of Paradise 90
      Non-Paradisical Doctrines 93
      Touching the Ark of GOD 96
      A New Kingdom 105

Book of Life Afterward and Epilogue 107
  The Twelfth Sinew of Adamant 107

Introduction
What is the Covenant Ark?

For over 2,000 years now, the Ark of the Covenant has been missing. It is believed that the Ark was taken during the Babylonian sack of Jerusalem of about 586 BCE. After that, there isn't any known dependable record of its location.

What was the Ark of the Covenant in the first place? Here is what we know. It was the symbol of the LORD's presence and authority among HIS people. It was designed to guide GOD's children through spiritual battle. It also had some form of divine ability.

The Ark of the Covenant was said to be a wooden box overlaid with gold. It had a throne for GOD with two winged beings overshadowing it.

What wasn't openly known was that the ark has three main parts. They are called the Ark of the Covenant, the Ark of the Testimony, and the Ark of GOD. These are one. Outside of the veil, it is a physical box. That box was designed to help conceal its true meaning. Within the veil, it is a testimony designed to scientifically prove that the Bible came from GOD. It is an accurate, authentic document. The testimony was designed to mathematically and spiritually link doctrines into Source Frequency. The doctrine can pierce science and nature. It has an alignment that draws directly through creation to GOD.

Babylon had stolen the Ark with theoretic explanations. They overpowered the people's comprehension of what Hebrews had explained during those days. Basically, the Babylonian description of creation seemed more accurate to the people. Their depictions eventually included the theory of relativity, string theory, and the big bang theory. Here is Gnostic evidence of this being true.

(Gnostic The Tripartite Tractate) "They were stronger than them in the lust for power, for they were more honored than the first ones, who had been raised above them. Those had not humbled themselves. They thought about themselves that they were beings originating from themselves alone and were without a source. As they brought forth at first according to their own birth, the two orders assaulted one another, fighting for command because of their manner of being. As a result, they were submerged in forces and natures in accord with the condition of mutual assault, having lust for power and all other things of this sort. It is from these that the vain love of glory draws all of them to the desire of the lust for power, while none of them has the exalted thought nor acknowledges it."

CODEX I Translated by Harold W. Attridge and Dieter Mueller Selection made from James M. Robinson, ed., The Nag Hammadi Library, revised edition. HarperCollins, San Francisco, 1990.

Babylon also had power by force, which seemed honorable to the people. Because of this, they had taken the third portion, the Ark of GOD. That will be explained in chapter 5 of this book.

The systems of books 7 and 8 of this series are tuned to source frequency. GOD's covenant is within them. The ark has the keys of the authority to do this. With it, the doctrinal structures were aligned correctly. This Ark is important to all religions. The Lamb's Book of Life has a platform grafted into GOD.

Here are several Gnostic verses that explain the source and why it aligns with Book 7, which is the Book of the Living.

(Gnostic Gospel of Truth) "They were glorified and they gave glory. In their heart, the living book of the Living was manifest, the book which was written in the thought and in the mind of the Father and, from before the foundation of the All, is in that incomprehensible part of him. This is the book which no one found possible to take, since it was reserved for him who will take it and be slain. No one was able to be manifest from those who believed in salvation as long as that book had not appeared. For this reason, the compassionate, faithful Jesus was patient in his sufferings until he took that book, since he knew that his death meant life for many. Just as in the case of a will which has not yet been opened, for the fortune of the deceased master of the house is hidden, so also in the case of the All which had been hidden as long as the Father of the All was invisible and unique in himself, in whom every space has its source. For this reason Jesus appeared. He took that book as his own. He was nailed to a cross. He affixed the edict of the Father to the cross. Oh, such great teaching! He abases himself even unto death, though he is clothed in eternal life. Having divested himself of these perishable rags, he clothed himself in incorruptibility, which no one could possibly take from him. Having entered into the empty territory of fears, he passed before those who were stripped by forgetfulness, being both knowledge and perfection, proclaiming the things that are in the heart of the Father, so that he became the wisdom of those who have received instruction. But those who are to be taught, the living who are inscribed in the book of the living, learn for themselves, receiving instructions from the Father, turning to him again. Since the perfection of the All is in the Father, it is necessary for the All to ascend to him."

(Gnostic The Gospel of Truth) "But they themselves are the truth; and the Father is within them, and they are in the Father, being perfect, being undivided in the truly good one, being in no way deficient in anything, but they are set at rest, refreshed in the Spirit. And they will heed their root. They will be concerned with those (things) in which he will find his root, and not suffer loss to his soul. This is the place of the blessed; this is their place.

CODEX I Translated by Harold W. Attridge and George W. MacRae Selection made from James M. Robinson, ed., The Nag Hammadi Library, revised edition. HarperCollins, San Francisco, 1990.

(Gnostic The Tripartite Tractate) "For the face of the copy normally takes its beauty from that of which it is a copy. They thought of themselves that they are beings existing by themselves and are without a source, since they do not see anything else existing before them."

CODEX I Translated by Harold W. Attridge and Dieter Mueller Selection made from James M. Robinson, ed., The Nag Hammadi Library, revised edition. HarperCollins, San Francisco, 1990.

(Gnostic The Tripartite Tractate) "For, thus they were honored in every place by him, being pure, from the countenance of the one who appointed them, and they were established: paradises and kingdoms and rests and promises and multitudes of servants of his will, and though they are lords of dominions, they are set beneath the one who is lord, the one who appointed them. After he listened to him in this way, properly, about the lights, which are the source and the system, he set them over the beauty of the things below."

CODEX I Translated by Harold W. Attridge and Dieter Mueller Selection made from James M. Robinson, ed., The Nag Hammadi Library, revised edition. HarperCollins, San Francisco, 1990.

(Gnostic Eugnostos the Blessed) "He is imperishably blessed. He is called 'Father of the Universe'. Before anything is visible among those that are visible, the majesty and the authorities that are in him, he embraces the totalities of the totalities, and nothing embraces him. For he is all mind, thought and reflecting, considering, rationality and power. They all are equal powers. They are the sources of the totalities."

(Gnostic Allogenes) "He is perfect, and he is greater than perfect, and he is blessed. He is always One and he exists in them all, being ineffable, unnameable, being One who exists through them all - he whom, should one discern him, one would not desire anything that exists before him among those that possess existence, for he is the source from which they were all emitted. He is prior to perfection. He was prior to every divinity, and he is prior to every blessedness, since he provides for every power."

(Gnostic Allogenes) "Since it is impossible for the individuals to comprehend the Universal One situated in the place that is higher than perfect, they apprehend by means of a First Thought - not as Being alone, but it is along with the latency of Existence that he confers Being. He provides everything for himself, since it is he who shall come to be when he recognizes himself. And he is One who subsists as a cause and source of Being, and an immaterial material and an innumerable number and a formless form and a shapeless shape and a powerlessness and a power and an insubstantial substance and a motionless motion and an inactive activity."

CODEX XI Translated by John D. Turner and Orval S. Wintermute Selection made from James M. Robinson, ed., The Nag Hammadi Library, revised edition. HarperCollins, San Francisco, 1990.

(Gnostic Trimorphic Protennoia) "It (the Word) is a hidden Light, bearing a fruit of life, pouring forth a living water from the invisible, unpolluted, immeasurable spring, that is, the unreproducible Voice of the glory of the Mother, the glory of the offspring of God; a male virgin by virtue of a hidden Intellect, that is, the Silence hidden from the All, being unreproducible, an immeasurable Light, the source of the All, the root of the entire Aeon. It is the foundation that supports every movement of the Aeons that belong to the mighty glory. It is the foundation of every foundation. It is the breath of the powers. It is the eye of the three permanences, which exist as Voice by virtue of Thought. And it is a Word by virtue of Speech; it was sent to illumine those who dwell in the darkness. Now behold! I will reveal to you my mysteries, since you are my fellow brethren, and you shall know them all.

CODEX XIII Translated by John D. Turner Selection made from James M. Robinson, ed., The Nag Hammadi Library, revised edition. HarperCollins, San Francisco, 1990.

(Gnostic A Valentinian Exposition) "He is a spring. He is one who appears in Silence, and he is Mind of the All dwelling secondarily with Life. For he is the projector of the All and the very hypostasis of the Father, that is, he is the Thought and his descent below."

CODEX XI Translated by John D. Turner Selection made from James M. Robinson, ed., The Nag Hammadi Library, revised edition. HarperCollins, San Francisco, 1990.

This book will reveal what the Hebrews and Gnostics were speaking about. You will learn how to align doctrines to Source Frequency. And you will find that the Bible is still far more accurate than all of Babylon's theories. Enjoy scientific proof that the Bible really came from GOD.

Chapter 1
Source Frequency of Creation

Twid oblivion
Before time's before
Time had no motion
Nor was in without
It wasn't was not in nor out
Without withouted without withouting
That unfoud, was founded undoubting
Something had to have, had not, hadn't, not have, have happened
In a way that didn't come about
By uhappeningly happening somehow though it did
When creation was created, which made you sacred

Locating the Ark of the Testimony

Through the years, a growing number of scholars have considered the Bible to be a mere manmade fabrication. Researchers have studied ceaselessly. Using symbols of educational rank to justify status of opinion, many were drawn away from belief. Faith in the unknown was being exchanged for trust in science.

There is a good reason for them not seeing the truth. The Bible was designed like the tabernacle of Moses and the temple of Solomon. There are several rooms in the Bible. These rooms are behind veils. Outside the temple was where the scholars were treading. There they only saw the words. The Bible resisted their unbelief.

Inside the temple are three main rooms: first the vestibule, second the sanctuary, and third the most holy place. Each room has a veil level. Anyone who has real spiritual faith, along with the spirit of GOD, can enter the vestibule. There they begin to see certain veils, allowing comprehension of what the Bible means. Next, we have the spiritually advanced and founders of the religions. They are able to enter the sanctuary. There they open veils allowing them to align their religions. In the sanctuary, people have more in-depth meanings revealed to them. Finally, GOD chooses who can enter the most holy place, also known as the holy of holies. In that room we find the Ark of the Testimony.

Here is a Gnostic reference about the third room of the temple.

(Gnostic Gospel of Philip) "At the present time we have the manifest things of creation. We say, "The strong who are held in high regard are great people. And the weak who are despised are the obscure." Contrast the manifest things that have been called truth: they are weak and despised, while the hidden things are strong and held in high regard. The mysteries of the truth are revealed, though in type and image. The bridal chamber, however, remains hidden. It is the holy in the holy. The veil at first concealed how God controlled the creation, but when the veil is rent and the things inside are revealed, this house will be left desolate, or rather will be [destroyed]. And the whole (inferior) godhead will flee [from] here but not into the holies [of the] holies, for it will not be able to mix with the unmixed [light] and the [flawless] fulness, but will be under the wings of the cross [and under] its arms. This ark will be [their] salvation when the flood of water surges over them. If some belong to the order of the priesthood they will be able to go within the veil with the high priest. For this reason, the veil was not rent at the top only, since it would have been open only to those above; nor was it rent at the bottom only, since it would have been revealed only to those below. But it was rent from top to bottom. Those above opened to us the things below, in order that we may go in to the secret of the truth. This truly is what is held in high regard (and) what is strong! But we shall go in there by means of lowly types and forms of weakness. They are lowly indeed when compared with the perfect glory. There is glory which surpasses glory. There is power which surpasses power. Therefore the perfect things have opened to us, together with the hidden things of truth. The holies of the holies were revealed, and the bridal chamber invited us in."

CODEX II Translated by Wesley W. Isenberg Selection made from James M. Robinson, ed., The Nag Hammadi Library, revised edition. HarperCollins, San Francisco, 1990.

As they spoke in the Gnostics, the veil concealed how GOD controlled the creation. (Hebrews 9:3) That is what the Ark of the Testimony is. The Ark is an unveiled document with scientific proof that the Bible came from GOD. Many researchers have come to believe that scientists are far more advanced than the Bible. The truth is the opposite. Scientists won't ever be ahead of GOD.

Next, we must locate the Ark and the bridal chamber.

(Gnostic Gospel of Philip) "Go into your chamber and shut the door behind you, and pray to your Father who is in secret" (Mt 6:6), the one who is within them all. But that which is within them all is the fullness. Beyond it, there is nothing else within it. This is that of which they say, "That which is above them". Before Christ, some came from a place they were no longer able to enter, and they went where they were no longer able to come out. Then Christ came. Those who went in, he brought out, and those who went out, he brought in. When Eve was still with Adam, death did not exist. When she was separated from him, death came into being. If he enters again and attains his former self, death will be no more."

CODEX II Translated by Wesley W. Isenberg Selection made from James M. Robinson, ed., The Nag Hammadi Library, revised edition. HarperCollins, San Francisco, 1990.

What we see in the Gospel of Philip is that the bridal chamber will bring Adam and Eve back together. This returns life to the doctrines. To reunite Adam and Eve, we must begin at the beginning. (Revelation 1:8) The Ark of the Testimony begins with the first two chapters of the Bible.

Universal Comprehension

Here is where mathematicians and physicists can begin to grasp the reality of scripture. Genesis chapters one and two explain the creation. It is written in dichotomies. These opposites are used to reveal the path in which GOD formed existence. The coming of creation is expressed with an abstract number system.

Human maticians had attempted to see a creation procession beginning from 1 or 0. That always leads to a dead end. (Tao Te Ching 2:2, 14:2-3) By dichotomies and contraries, the universal laws of physics were created. The dichotomy of something is the lack of something. Nothing doesn't have an opposite. Nothing is neither something nor the lack thereof. (Tao Te Ching 40:1-2)

Creation couldn't begin with a positive nor a number because both are impassable. Positives only make positives. They don't lead to dichotomies. Numbers alone have no mathematics to equate change. Numbers don't have plus, multiply, subtract, or divide. Numbers are numbers. If mathematics were not yet created, then numbers couldn't be used together.

The dichotomized mathematics in Genesis are for a flawless number system. It is a truth, not a theory. To understand this, we begin at the beginning before beginnings began. There were no laws of physics, and the only rule was GOD.

Unlike computer binaries, a number system for creation cannot begin with 1 or 0. It must begin with GOD. Without GOD, creation theories enter a dead end. Let's begin with a symbol signifying the Creator. To make this simple, we will use the plus (+) symbol. This plus will symbolize the root of all existence. It is called the first positive of wisdom. (Matthew 19:26)

(Gnostic Gospel of Mary) "The Savior said "All natures, all formations, all creatures exist in and within one another, and they will be resolved again into their own roots. For the nature of matter is resolved into the (roots) of its nature alone. He who has ears to hear, let him hear." (Tao Te Ching Chapter 6)

The plus symbol, known as the first positive, signifies the LORD. Being that it is HE, it would align to knowledge. It holds a wisdom alignment because She, wisdom, is ONE with HIM. (Tao Te Ching 55:3-4) This plus symbol represents that which we don't fully understand. It represents Source Frequency.

(Gnostic Apocryphon of John) "For we know not the ineffable things, and we do not understand what is immeasurable, except for him who came forth from him, namely (from) the Father. For it is he who told it to us alone. For it is he who looks at himself in his light which surrounds him, namely the spring of the water of life. And it is he who gives to all the aeons and in every way, (and) who gazes upon his image which he sees in the spring of the Spirit. It is he who puts his desire in his water-light which is in the spring of the pure light-water which surrounds him."

CODEX II Translated by Frederik Wisse Selection made from James M. Robinson, ed., The Nag Hammadi Library, revised edition. HarperCollins, San Francisco, 1990.

First Negative of Wisdom

If a number system for creation began with 1, what would be next? Would you add another 1 to it? You couldn't, because addition wouldn't exist yet. A positive number doesn't lead to dichotomies? If you did have addition, you could add numbers together forever, and there still wouldn't be any opposites. Without mathematics or numbers is where we begin.

At the starting point, the ONE thing that existed was LOVE. (1 John 4:8) Creation as we know it flowed from GOD like a spring.

The way that we can first comprehend creation is in words such as lack or minus.

Consider this: we view creation from this side of existence. To understand how we came to be, first we must subtract that which we have and all that we are. Subtract your home, your life, your children, the earth, the galaxy, the universe, and all of existence. What then would be is the absence of all that there is. That would bring us the first symbol of creation as a minus (-). This minus symbol represents lacking everything in existence. It is the first negative of wisdom. (Proverbs 1:20)

The first negative isn't a number because the concept of numbers didn't exist. All that would be wouldn't have been, and therefore minus everything that exists today. It symbolizes a figurative negative property instead of a numerical value.

At that point, zero didn't happen because zero would mean nothing. Naught can anything be created from zero. Zero has no abstract or opposite polarity; zero is only nothing. GOD alone existed, and therefore nothing couldn't have been.

Second Negative of the Creation of Understanding

In the beginning, there was only the figurative concept of lacking all that there is today. The only thing that could exist was the property of negative, meaning minus existence.

Once we grasp that figurative reality, we find that creation couldn't have been lacking anything. This is because absence had not yet been created. At that point, creation became absent of the ability to lack. Everything that you had subtracted in the previous step couldn't have been subtracted because you also have to subtract subtraction.

The universe lacked the negative property needed in order to lack existence. (Tao Te Ching 23:3) A diagram of Creation at that point would have two minus symbols (- -). The second subtraction is the second negative of the creation of understanding. (Proverbs 9:4)

There weren't two negatives in count because numbers hadn't been created. Creation was still only figuratively figurative as properties.

Third Negative of the Creation of Instruction

At this point, it should be obvious that the possibility of creation having subtracted subtraction would need to be subtracted as well. Everything cannot have been subtracted without having lacked the lack of lacking.

Existence would be a mere figuration that was figuratively figurative. To represent this, we would use three subtraction symbols like this (- - -). The third subtraction is the third negative of the creation of instruction. (Proverbs 4:13)

We find clear evidence of the first three negatives in the Tao Te Ching. (Tao Te Ching 14:1, 42:1, 62:3)

These creation mathematics are also in the Qur'an, where they speak of how this truth provides no hiding place for deniers. After the people have denied the various religions and the Bible, they are then instructed to return to them. The three negatives here are called three columns.

(Winds Sent Forth Surah 77:28-37) "Woe, on that Day, to those who denied the truth! They will be told, 'Go to that which you used to deny! Go to a shadow of smoke!' It rises in three columns; no shade does it give, nor relief from the flame; it shoots out sparks as large as tree-trunks' and as bright as copper. Woe, on that Day, to those who denied the truth! On that Day they will be speechless, and they will be given no chance to offer any excuses. Woe, on that Day, to those who denied the truth! [They will be told], 'This is the Day of Decision: We have gathered you and earlier generations. If you have any plots against Me, try them now' Woe, on that Day, to those who denied the truth!" (From the M. A. S. Abdel Haleem edition.)

These three negatives are also called fire, smoke, and brimstone. They are as a plague raining down from heaven upon scientific theories that were used to deny GOD. (Revelation 9:17-18) Though these can be seen as a plague to deniers, they are a blessing to believers. The walls of Jericho signify the barrier between religions and science and religion.

In the Bible, the three negatives are considered feminine and are called wisdom, instruction, and understanding. (Proverbs 4:13, 7:4, 8:1-5, 9:4-6) (Tao Te Ching 6:1) Them being feminine can be used to signify GOD. The first positive is masculine and can be used to signify the LORD. They have always been together as ONE. (Proverbs 8:22-26) Because GOD is ONE, there has long been debate on whether or not the LORD has a wife. The fact is that the LORD and GOD are not married in the realization that they are ONE perfect unification. (Proverbs 24:14)

Second Positive of Knowledge

That figurative property of negative, or lacking as it is being explained, would go on forevermore. A diagram of this would be infinite minus symbols to the infinite power, infinitely.

Three subtractions would only be followed by a fourth negative, and that would go on forever. Yet let us just add this fourth and final subtraction to represent an eternal continuance of negatives. Creation before existence is symbolized like this (- - - -).

These four minus symbols represent three similar negatives with a fourth that is somewhat different. There are a couple of artifacts that have aligned like this. The cross of Jesus Christ has three sides similar in length, with a fourth side elongated. The Shroud of Turin was woven with a three-over-one herringbone weave. These aren't coincidences. These four negatives are sometimes known as birds.

(The Cow Surah 2:260) "And when Abraham said, 'My Lord, show me how You give life to the dead,' He said, 'Do you not believe, then?' 'Yes,' said Abraham, 'but just to put my heart at rest.' So God said, 'Take four birds and train them to come back to you. Then place them on separate hilltops," call them back, and they will come flying to you: know that God is all powerful and wise.'" (From the M. A. S. Abdel Haleem edition.)

From that point, there must have been a way for creation to have moved forward.

Let us look a little closer at the masculine and feminine. Masculine is symbolized by the first positive. The masculine knowledge of love creates by organizing into increments. Because of this, we get infinite negative properties.

Feminine is signified by the negatives. GOD is emotion, will, and time in the feminine aspect. GOD is love. The wisdom of love wills and wants to pull things together. Wisdom's love would sanctify the negatives in spiritual unity. Love pulls the properties of minus-minus-minus together and subtracts the lack of existence.

Masculine made many increments, while feminine brought them together as one universal body.

Lacking becomes the will to attract. When we as people lack something, we begin seeking. It is automatic that the absence of something initiates drawing or pulling inward. (John 12:32) Basically, creation filled the lack of existence with the want of existence.

(Gnostic The Tripartite Tractate) "Therefore, in the song of glorification and in the power of the unity of him from whom they have come, they were drawn into a mingling and a combination and a unity with one another. They offered glory worthy of the Father from the pleromatic congregation, which is a single representation although many, because it was brought forth as a glory for the single one and because they came forth toward the one who is himself the Totalities."

(Gnostic The Tripartite Tractate) "It appears and draws them to love the exalted one, to proclaim these things as pertaining to a unity."

CODEX I Translated by Harold W. Attridge and Dieter Mueller Selection made from James M. Robinson, ed., The Nag Hammadi Library, revised edition. HarperCollins, San Francisco, 1990.

Once you can comprehend that information, it is time to move forward with the creation process.

Feminine negatives must transform into masculine positives.

(Gnostic Gospel of Thomas 114:2-3) "Jesus said: "Look, I will draw her in so as to make her male, so that she too may become a living male spirit, similar to you." (3) (But I say to you): "Every woman who makes herself male will enter the kingdom of heaven."

CODEX II Translated by Stephen Patterson and Marvin Meyer Selection from Robert J. Miller, ed., The Complete Gospels: Annotated Scholars Version. (Polebridge Press, 1992, 1994).

Here is the process.

To comprehend being without creation is to recognize that the essence of the ability for creation to exist was already there. You cannot lack creation without creation being possible. The reflective ability was already with GOD before the beginning. (Proverbs 27:19) The LORD is ONE and isn't linearly timed. The LORD is before and after creation; therefore, creation can be seen in forward and reverse as pertaining to nonlinear time. To be without something means that there once was or could be that something. Time is universal in this aspect.

Subtraction was the first form of figurative mathematics. It was created before the beginning didn't even exist. Subtracting everything that you know only makes it appear again.

Consider how mathematics works. If you subtract a negative number from the same negative number, you would get zero. If you had two different negative numbers, you could subtract one from another to get a positive number.

Using three negatives, one can be subtracted from another to form a positive. A minus property was subtracted from a minus property. (Tao Te Ching: 14:1) The fourth negative symbol remains with the positive when the first three merge. (- - - = - +)

The First Day of Creation

This second positive property was created on the first day of Biblical Creation. (Genesis 1:3-5) In this description, darkness was negative, and light was positive. Writing it would look like negative and positive symbols next to each other (- +). Infinite negatives allow for a negative to remain.

The first plus symbol signifies the Creator. The second positive is the second positive of divine knowledge. Let there be light. (2 Corinthians 4:6) That was the first day of universal creation.

Multiplicative of Understanding

Once the second positive was created as the fourth figurative property, there were now figurative numbers. That is why the Book of Numbers is fourth in the Bible. Infinite negatives would allow for unlimited positives to be formed. Only if the negatives were in count, to make a positive, would numbers exist. This is because infinity ain't a number.

GOD had separated the two properties in dichotomy. This is when the impossible zero became possible. Nothing, as a figurative numeric concept, was fourth. Zero couldn't have been until after both negative and positive existed. Something had to exist before a concept could be neither one nor the other. The zero represents the neither property. Zero was placed between negative and positive. (- 0 +)

The Second Day of Creation

On the second day of creation, GOD made a firmament to distinguish the difference between the waters. (Genesis 1:6-8) The waters signify spirits. There would eventually be spirits in a physical realm below and in the spiritual realm of GOD above. The firmament was placed in the midst of the waters.

This was possible because positive and negative were dichotomized. It allowed the ability for two things to exist through contrast. The spiritual realm is the negative, and the physical realm is the positive. Spiritual, being infinite like GOD, was able to subtract and merge to create the possibility of a physical plane.

The firmament is also called Heaven. It is the place where GOD personally connects with that which was created. The firmament is a perfect unity of spiritual and physical. In the spiritual realm, positive is the entity of the possibility of physical. This made the opportunity for two forms of creation. The firmament is where the two developments are perfectly united.

Here is a geometric visual of creation after the second day.

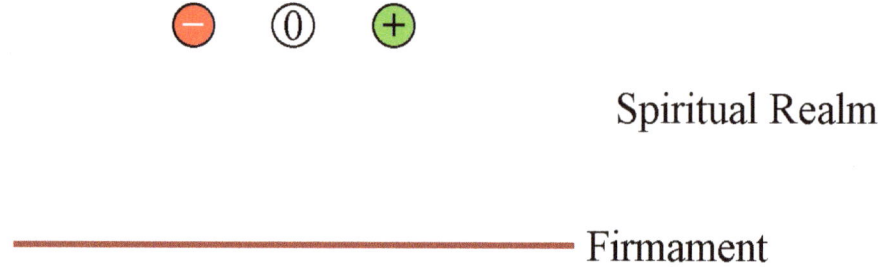

GOD had already made the dichotomy of minus and plus. You may now notice that with subtraction and addition, GOD had multiplied the possible properties by two. This means that multiplication was formed as soon as positive was created. On the third day, the mathematical possibilities became three.

Mathematical division, on the other hand, hadn't yet been created. The reason was because plus didn't actually divide from minus. Zero didn't divide anything; it was neither one nor the other, as if nothing. Minus basically subtracted itself like an implosion. The positive is a dual merged negative entity. See how a plus symbol is made from two minus symbols. Positive isn't divided from negative; it is merged negatives.

Third Day of Creation

On the third day, the seas were gathered, and the earth appeared. (Genesis 1:9-10) Seas signify negative properties. The earth signifies positive properties. The gathering of seas represents the mathematical possibility to multiply negative properties. The appearance of earth represents the mathematical possibility to multiply positive properties.

Gravity of Equity and Pressure of Perception

GOD had now created the two forms of multiplication, including multiplication of positive properties (appearing of earth) and multiplication of negative properties (gathering of seas).

The herb seeds signify multiplication yielding each according to its kind, either positive or negative. Negative was multiplied to create gravity. Positive was multiplied to create pressure. (Tao Te Ching 26:1)

Gravity is represented by the grass, and pressure is represented by the fruit trees. Each multiplication is according to its own kind as an increment. The gravity pulls the fruit off the fruit tree. It falls and lands on the grass with pressure. (Genesis 1:11-13)

Whatever multiplies on the positive side is of the Father and the Spirit. He incorporates the laws of physics and knowledge. Whatever multiplies on the negative side is of the Mother and the Holy Ghost. She incorporates will, emotion, and wisdom. Within HIM, SHE finds HER throne. (Tao Te Ching 55:3)

Here is a geometric visual of creation after the third day.

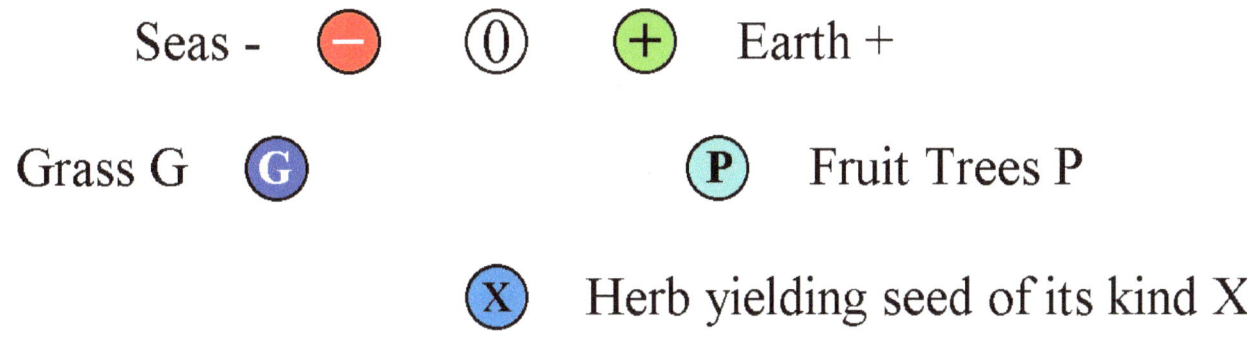

Fourth Day of Creation

GOD then made lights in the firmament. The greater and lesser lights are the sun and the moon. They signify parents. The stars represent children. This means that in the firmament, known as Heaven, there was a place prepared for us before our creation on earth. (Genesis 1:14-19) Imagine Heaven like a perfected unity of spiritual and physical. It is a place where anything is possible. Then consider the fact that Heaven is even beyond both realms. It is beyond spiritual. Heaven is very real and far greater than we have learned to imagine.

The lights also signify a plan for two realms. The greater light to govern the day is the spiritual realm of femininity and emotion. The lesser light to govern the night is the physical realm of masculinity and logic. As a mirror, in the physical realm, the masculine is considered the greater light, which rules the day. When a woman wears her man, she then wears the greater light. This can be seen in Revelation chapter 12, where the woman clothed with the sun has the moon under her feet. The sun that she wears is her husband. He is also clothed with her (the moon) helping give birth.

Along with that, the sun and moon represent a Father and Mother in Heaven, while the stars signify angels. (Revelation 1:20)

In these picture graphs, the firmament is the dark red line. Above the line is the spiritual realm where creation begins. Below the line is the physical realm.

Here is a geometric visual of creation after the fourth day.

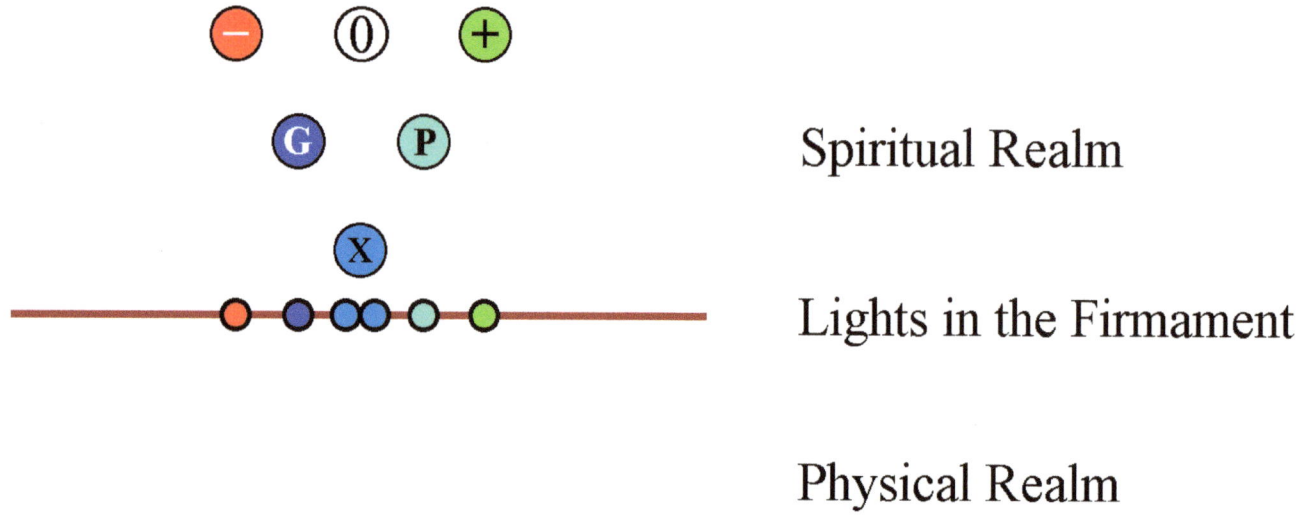

Division

Once positive and negative were multiplied, division was created. Multiplication had been divided into two groups. Multiplication of negative was one form. Multiplication of positive was the other form. This division of property usage was now available.

Volume of Discretion and Density of Counsel – Fifth Day of Creation

On the fifth day, GOD made the sea creatures and the birds. (Genesis 1:20-23) To do this, the newly created mathematical property of division was used with gravity and pressure.

The sea signifies the spiritual property and realm of negative (emotion). The earth signifies the physical(ity) property and realm of positive (logic). If the fish go into the sea, then the fish are the masculine. If the birds are on dry land and in the air, then they are feminine. And here is why.

(Gnostic Gospel of Philip) "Then Christ came. Those who went in, he brought out, and those who went out, he brought in. When Eve was still with Adam, death did not exist. When she was separated from him, death came into being. If he enters again and attains his former self, death will be no more."

CODEX II Translated by Wesley W. Isenberg Selection made from James M. Robinson, ed., The Nag Hammadi Library, revised edition. HarperCollins, San Francisco, 1990.

The bird being the feminine property means that she is connected to him in the Holy Spirit. (Luke 3:22) This connection is mirrored with the sea creatures. It is brought about by the factor of division interacting with both positive and negative. The masculine fish were divided into the sea (her). The feminine birds, like Jonah the Dove, were divided into the masculine air (him). (Tao Te Ching 36:2) This means that although positive is primarily seen as the masculine aspect, it, like negative, has a gender duality. More on Jonah's identity is revealed in book 7 of this series.

A division of gravity is volume, symbolized by the birds. A division of pressure is density, symbolized by the sea creatures. On the fifth day, this mathematics happened in the spiritual realm. Everything is created in the feminine spiritual realm before it is created in the masculine physical realm. That which the physical plane receives comes from the spiritual plane.

GOD also allowed the birds to fly above the earth across the face of the firmament. This means that the feminine negatives (birds) of creation passed through the firmament into the masculine physical realm. This is similar to the dove landing on Jesus in bodily (physical) form.

The spiritual realm is the feminine (negative), where masculinity (positive) is her physicality. The physical realm is masculine (positive), where femininity (negative) is his physicality. To create a duality of masculine and feminine coming back together, spirituality became masculine within the physical realm.

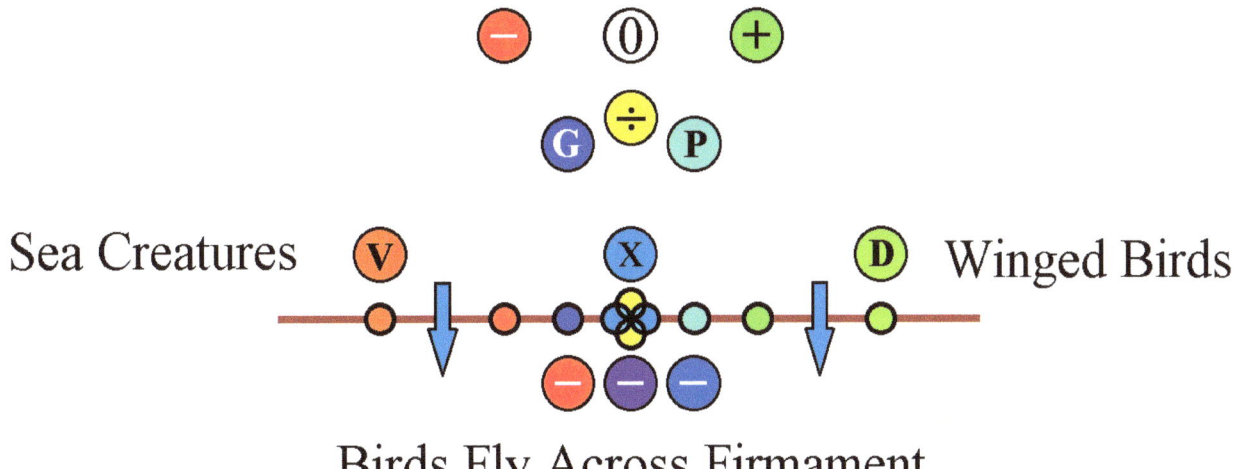

That was the fifth day of universal creation.

Sixth Day of Creation

On the sixth day, GOD let the earth (physical realm) bring forth every creature according to its kind. First, cattle, creeping thing, and beast are mentioned in order. Next, the order of beast is moved before cattle and everything that creeps. Beast shifts because it is divided. (Genesis 1:24-25) The physical realm receives the same mathematics and/or processes as the spiritual realm.

Man was then created in the image of GOD as male and female. This relates to the form of GOD represented by the first positive symbol. The first plus (masculine) symbol is called the first positive of wisdom, yet wisdom signifies feminine or negative. First positive of wisdom means feminine and masculine as one. The LORD GOD is ONE, with both genders completely united. (Genesis 1:26-31) During this time, Eve was still within Adam. They were created in the likeness of GOD.

This part of Genesis is about man being given dominion over that which GOD had created. Man being mentioned signifies that this second realm is the physical realm. Everything on the sixth day happened on the physical plane.

The earth brought forth the living creature according to its kind.

Cattle means negative (-).
Creeping thing means positive (+).
Herbs that yield seed means multiplication (x).
Green herbs for food means gravity (grass is herbaceous) (G).
Trees whose fruit yields seed means pressure (P).
Beast means division ($\div$).
Fish means volume (V).
Birds mean density (D).

Division (beast) is called divisive of instruction. Instruction (division) was called beast because beast was instructing humanity. Division is now re-sourced to GOD. There are many small hints and pieces scattered about the Gnostic books.

(Gnostic On the Origin of the World) "An androgynous human being was produced, whom the Greeks call Hermaphrodites; and whose mother the Hebrews call Eve of Life (Zoe), namely, the female instructor of life. Her offspring is the creature that is lord. Afterwards, the authorities called it "Beast", so that it might lead astray their modelled creatures. The interpretation of "the beast" is "the instructor". For it was found to be the wisest of all beings."

(Gnostic On the Origin of the World) "And regarding these, the holy voice said, "Multiply and improve! Be lord over all creatures."

"The Untitled Text" CODEX XIII Translated by Hans-Gebhard Bethge and Bentley Layton Selection made from James M. Robinson, ed., The Nag Hammadi Library, revised edition. HarperCollins, San Francisco, 1990.

On the sixth day, GOD created mathematical properties on earth (physical realm) like a mirror to the spiritual realm. The feminine spiritual realm is based on emotion, where mathematics is figurative. The masculine physical realm is based on mathematics, where emotion is figurative.

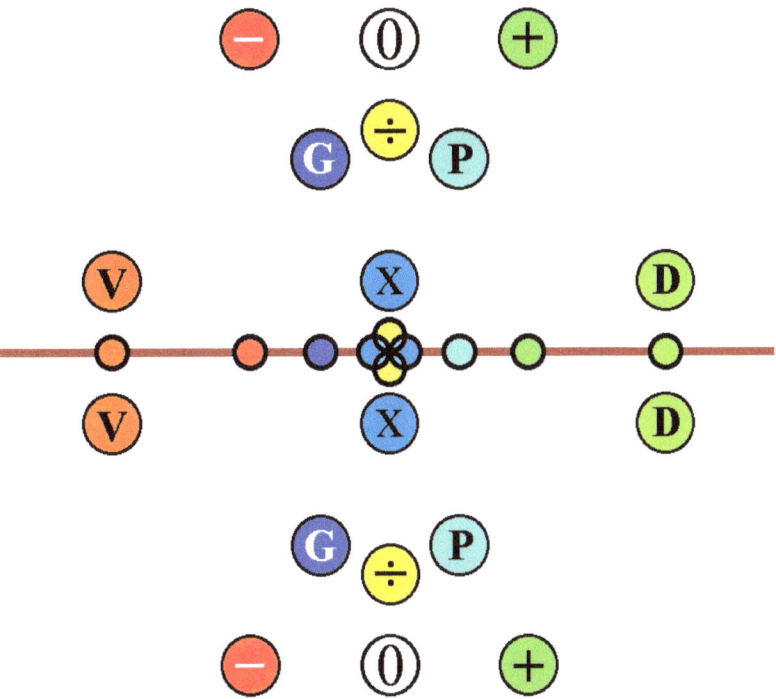

That was the sixth day of universal creation.

Weight of Prudence – Seventh Day of Creation

On the seventh day of creation, all that GOD had created was sanctified. The entirety of mathematics came together as one. To do this, volume and density were multiplied to make weight. This sanctified the mathematics, and they all worked together in unity.

To symbolize the sanctification, a dark red line was drawn in a W connecting the universal mathematics. Zero is left out of the W because zero is nothing. Division isn't in the W because it represents instruction which is transferred. The W is mirrored within the two realms of spiritual and physical. The two polarities signify dual triangle waves. They would eventually pass through each other as Adam and Eve get closer to each other.

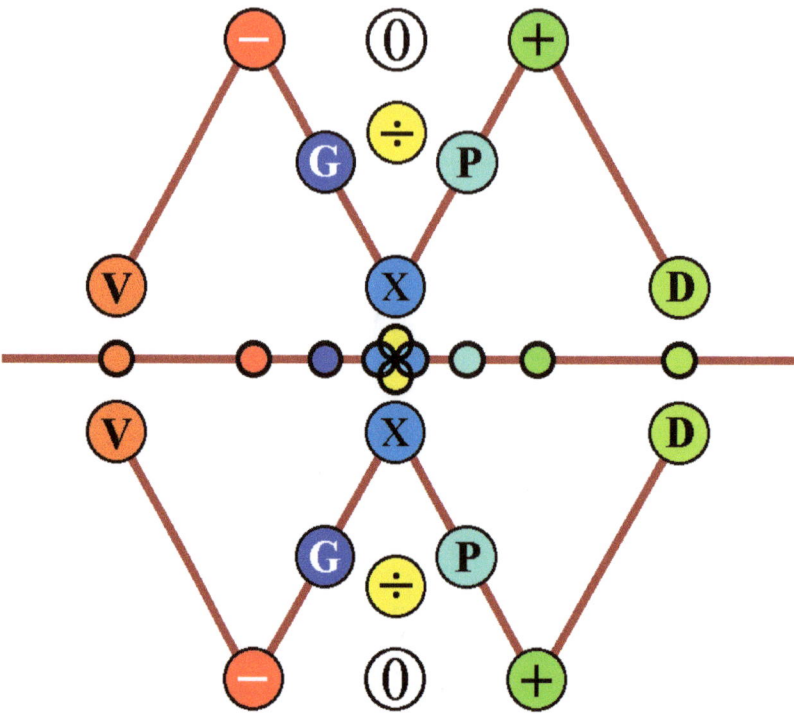

We notice that although man was created during the sixth day, Adam was created on or after the seventh day. In this veil, man in the first six days wasn't referring to a person. Man on the sixth day refers to masculine physicality and the physical realm.

The mathematics were first formed. From this mathematics, matter (dust) was made. Adam was created from the dust (or clay) of the earth (physical realm). The breath of life was then breathed into him, and he received Spirit. (The Believers Surah 23:12) The lady in the physical realm was still within Adam, in a spiritual form, as a rib or a bird of clay. (The Family of 'Imran Surah 3:49-50)

The physical realm is masculine. Feminine, from the spiritual realm, entered the physical realm by masculine. Eve was brought here through Adam. (Sirach 42:24-25) (Tao Te Ching 29:2, 43:1-2) She was within him and passed through him to get here.

(Gnostic The Tripartite Tractate) "The first human being is a mixed formation, and a mixed creation, and a deposit of those of the left and those of the right, and a spiritual word whose attention is divided between each of the two substances from which he takes his being."

CODEX I Translated by Harold W. Attridge and Dieter Mueller Selection made from James M. Robinson, ed., The Nag Hammadi Library, revised edition. HarperCollins, San Francisco, 1990.

The physical realm is governed by mathematical laws of physics. The spiritual realm is governed by emotion. We cannot touch emotion the same way as a piece of wood, yet there are ways for the wood to stimulate emotions. We cannot feel mathematics the same way we feel happiness, yet happiness can add up and multiply. In the spiritual realm, mathematics is figurative. In the physical realm, emotion is figurative. These realms come together as one. All things are real.

What this chapter revealed is the realms that we know of GOD creating. The explanation has some similarities to the big bang theory, yet this isn't just a theory. It is the universal mathematics of rudimentary physics. GOD has done much more that we don't know about.

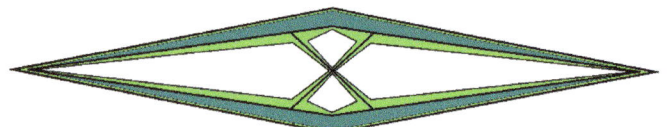

Chapter 2
Integrating the Ark

This chapter will reveal what the Ark of the Testimony is for. You will learn how doctrines are infused using the Ark with the Urim and Thummim.

The seven days of creation are set to reveal GOD's number system through abstract mathematics. Here is the simplified version. First was subtraction. Then subtraction subtracted subtraction from itself to make addition. The presence of subtraction and addition was a multiplication in properties. Now there was multiplication. Subtraction and addition were both multiplied to make gravity and pressure. By multiplying subtraction and addition, multiplication was divided into two. Now there was division. Gravity and pressure were divided to make volume and density. Finally, volume and density were multiplied to make weight.

Here are the mathematical stages during the seven days of creation. They are the meaning of the seven menorah candles. They are also called the seven pillars of wisdom. (Proverbs 9:1)

1: Subtraction
2: Addition
3: Multiplication
4: Gravity and Pressure
5: Division
6: Volume and Density
7: Weight

When we separate each mathematical property, there are nine. The eighth is density, and the ninth is weight. Density multiplied by volume reveals the weight. This is spoken of in the Gnostics.

(Gnostic The Discourse of the Eighth and the Ninth) ""My father, yesterday you promised me that you would bring my mind into the eighth and afterwards you would bring me into the ninth. You said that this is the order of the tradition." "My son, indeed this is the order."

(Gnostic The Discourse of the Eighth and the Ninth) "Therefore I say that they are immortal." "Your word is true; it has no refutation from now on. My father, begin the discourse on the eighth and the ninth, and include me also with my brothers." "Let us pray, my son, to the father of the universe, with your brothers who are my sons, that he may give the spirit of eloquence.""

(Gnostic The Discourse of the Eighth and the Ninth) "Rejoice over this! For already from them the power, which is light, is coming to us. For I see! I see indescribable depths. How shall I tell you, my son? [...] from the [...] the places. How shall I describe the universe? I am Mind, and I see another Mind, the one that moves the soul!"

(Gnostic The Discourse of the Eighth and the Ninth) "We have received this light. And I myself see this same vision in you. And I see the eighth, and the souls that are in it, and the angels singing a hymn to the ninth and its powers. And I see him who has the power of them all, creating those <that are> in the spirit.""

(Gnostic The Discourse of the Eighth and the Ninth) ""I will offer up the praise in my heart, as I pray to the end of the universe and the beginning of the beginning, to the object of man's quest, the immortal discovery, the begetter of light and truth, the sower of reason, the love of immortal life."

(Gnostic The Discourse of the Eighth and the Ninth) "Rather, by stages he advances and enters into the way of immortality. And thus he enters into the understanding of the eighth that reveals the ninth.""

CODEX VI Translated by James Brashler, Peter A. Dirkse, and Douglas M. Parrott Selection made from James M. Robinson, ed., The Nag Hammadi Library, revised edition. HarperCollins, San Francisco, 1990.

We can also view mathematical creation in nine stages. When searching these books, we find that the steps of creation are called the nine gates.

(Gnostic The Acts of Peter and the Twelve Apostles) "This is the name of my city, 'Nine Gates' Let us praise God as we are mindful that the tenth is the head."

CODEX VI Translated by Douglas M. Parrott and R. McL. Wilson Selection made from James M. Robinson, ed., The Nag Hammadi Library, revised edition. HarperCollins, San Francisco, 1990.

(Bhagavad Gita 5:13) "Those who renounce attachment in all their deeds live content in the "city of nine gates," the body, as its master. They are not driven to act, nor do they involve others in action." (Eknath Easwaran Edition)

The nine gates are the nine properties when considered as separate and individual. They are the meaning of the nine candles on the Hanukkah Menorah. The tenth gate, which is called the head, is the first positive that represents the Creator.

1: Subtraction (Wisdom)
2: Addition (Knowledge)
3: Multiplication (Understanding)
4: Gravity (Equity)
5: Pressure (Perception)
6: Division (Instruction)
7: Volume (Discretion)
8: Density (Counsel)
9: Weight (Prudence)
10: Creator (First Positive of Wisdom) (Revelation 22:13)

The Maya spoke of these tower stories as the Bolon Yokte' or Nine Strides. These nine stories are nine mathematical strides of manifestation. They are also like the nine months of the gestation of creation. This is the matrix of the mother's womb. (NKJV Isaiah 49:1)

This next page has a depiction of the nine processes (gates) of creation as GOD revealed to the ancients.

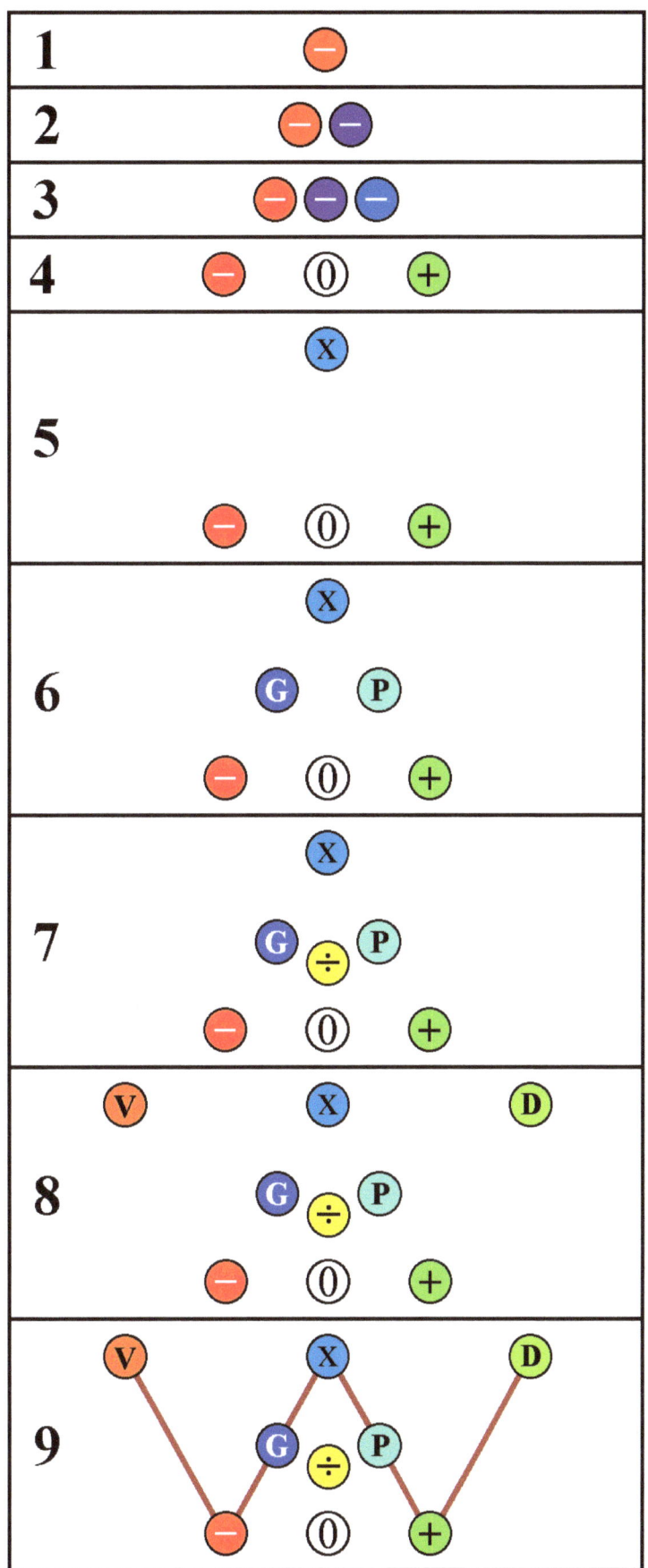

Once we understand the geometry of it all, we can begin to see something similar to algebra.

From the physical realm, we use algebra to find a solution by breaking questions down. An algebraic question begins with a mathematical line such as $(8x - 6)/8 + (6x - 4)/10 = (x + 12)/14$. Each step to solving the question breaks the line down. In the end, a single solution is found.

GOD's creation can be seen similarly in a mirror. GOD begins with the solution because GOD is the answer. From the solution, a functioning mathematical process is built. GOD can use infinite subtractive properties to create any number of positives. The process can be seen in the abstract, like this mathematical line. '(- -) (- -) (- -) = +3.'

If you were going to use spiritual mathematics to do algebra, it may look something like a tower of property alignments. This is what the Tao Te Ching meant by the next quoted verse. It was also part of the Kesh Temple design.

(Tao Te Ching 64:2) "The tree which fills the arms grew from the tiniest sprout; the tower of nine stories rose from a (small) heap of earth; the journey of a thousand li commenced with a single step." (James Legge Edition)

$$(-)$$
$$(--)$$
$$(---)$$
$$(- +)$$
$$(- 0 +)$$
$$(- x\ 0\ x +)$$
$$(- G\ x\ 0\ x\ P +)$$
$$(- G\ x\ V \div 0 \div D\ x\ P +)$$
$$(W - G\ x\ V \div 0 \div D\ x\ P + W)$$

As in algebra, you would then place numbers next to the property symbols to signify their numerical value. Creation could then be calculated.

Scriptural Merging Procedures

The mathematics of creation are used to write scriptures. They are aligned to Source Frequency. Designing doctrines using this formula connects them to the Source of Creation. This blueprint of creation isn't only on a physical level. These properties are spiritual. Standard mathematics are limited to direct physical numbers. This creation path is a physical abstract from the spiritual realm.

For these raw calculations to be used, they must be placed into the Urim and Thummim. Before that can be done, they are aligned to the color wheel with Noah's rainbow covenant. This next picture is made from two squares overlapped. It is called the Rub El Hizb or Star of Lakshmi. It properly places the creation mathematics into a color circle.

Here is how they enter.

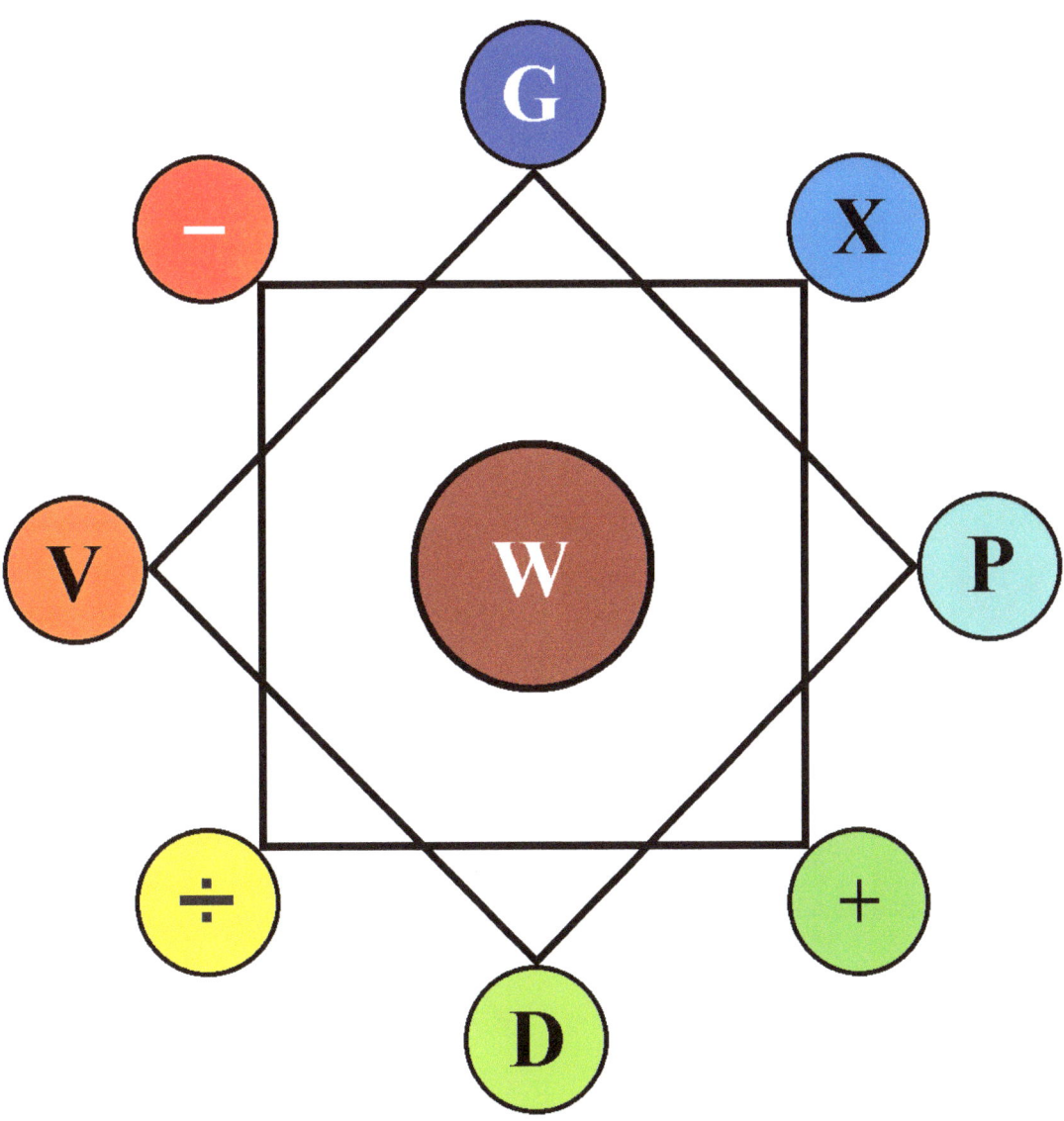

The Urim is a bow-shaped design. This sacred geometry is for properly aligning and using the universal mathematics. It has each of the nine fundamental properties of creation. It also includes positions for the three extra symbols of first positive, second negative, and third negative. The Urim is designed with three circles for prime colors. One of the circles is separated into two portions. The separation is for RGB and RYB models, which vary in prime colors yellow and green. Between the sections for yellow and green, the chartreuse is drawn out. In-depth directions on why and how the Urim is designed this way can be found in Book 7 of this series. The next diagram is a picture of the real Urim. You will see three circles, with one having two sections.

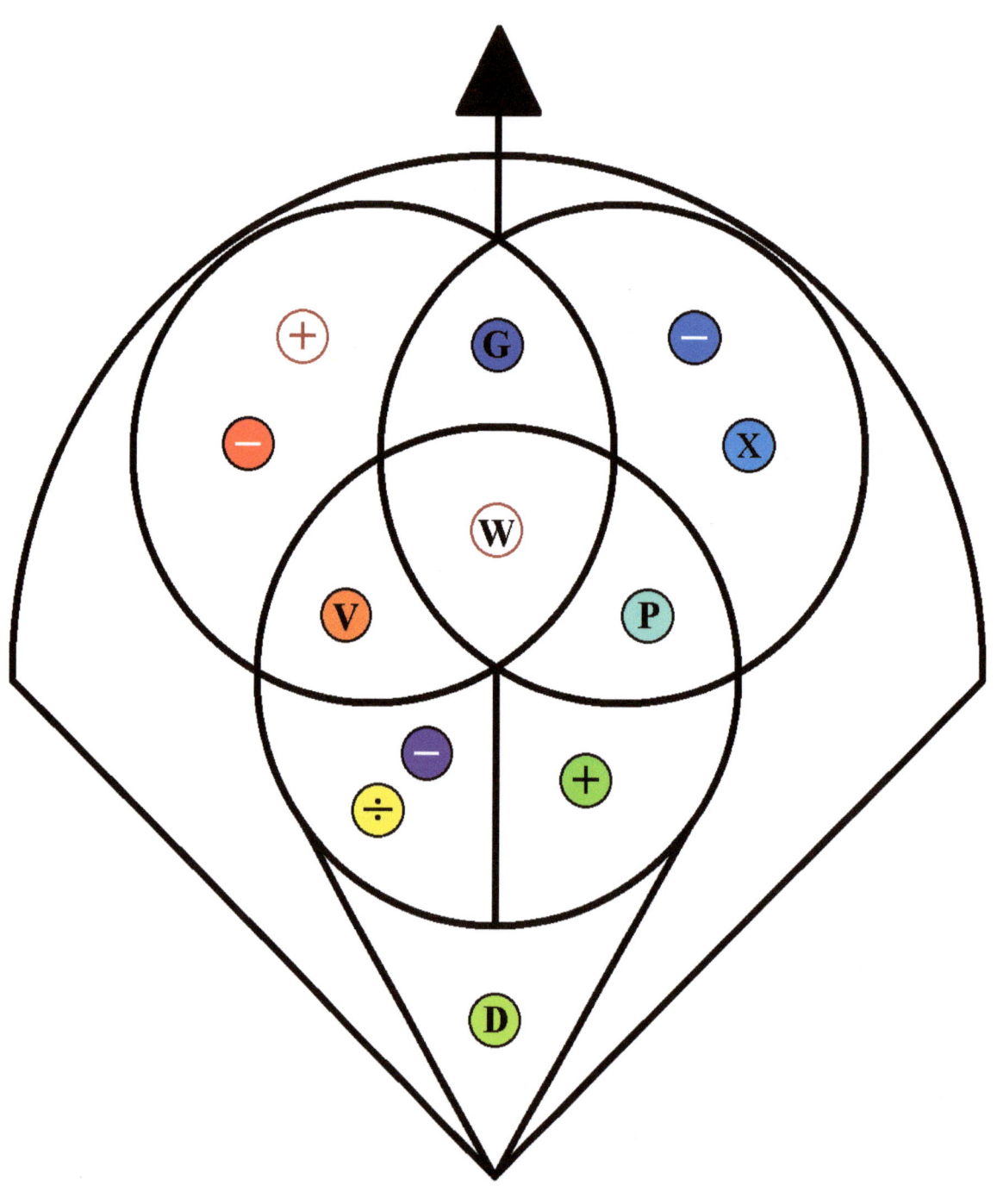

The Urim has seven areas surrounding a ninth, with an eighth section drawn out. Once this shape is correct, a perfect science can be inserted. The universal mathematics symbols are placed inside and aligned to a color wheel. The correct colors and musical notes are inserted. Information on how colors are aligned to notes can be found at this next website link.

https://www.endolith.com/wordpress/2010/09/15/a-mapping-between-musical-notes-and-colors/

The three main circles are aligned to atomics. The top left circle is for electron. The bottom circle is for proton. The top right circle is for neutron.

The Urim's psychology is based on three forms: emotional, cognitive, and social. They harmoniously enter the Urim as six psychological kinships. These are spoken of in Taoism. (Tao Te Ching 18:2) The six are emotion, cognition, social, social emotion, social cognitive, and cognitive emotion. When these kinships fell into disorder throughout the world, GOD sent loyal ministers to help the people.

There is also a precisely aligned set of nouns in each of the nine sections. Those words, such as wisdom and knowledge, are entered according to the directions in the biblical book of Proverbs.

Here is a list of the nine portions of the Urim, aligned to atomics, color, and psychology.

1: First Negative of Wisdom – Red – Electron – Emotion
2: Second Positive of Knowledge – Green – Proton – Cognition (cognitive knowledge)
3: Multiplicative of Understanding – Blue - Electron – Social
4: Gravity of Equity – Purple – Social Emotion
5: Pressure of Perception – Cyan – Social Cognitive
6: Divisive Instruction – Yellow – Cognition (cognitive instruction)
7: Volume of Discretion – Orange – Cognitive Emotion
8: Density of Counsel – Chartreuse – Cognition (cognitive counsel)
9: Weight of Prudence – White or Dark Red – Social/Emotional/Cognitive

You will find that in the Urim graph, the proton circle has three portions. These are for yellow, chartreuse, and green. That is from the three negatives of creation. The essence of negatives being three allowed for the proton (positive) circle to have three portions.

The Urim is an eight- and nine-point system. To bring it into a twelve-point system, we add in three extra symbols as seen in the Urim graph. First positive of wisdom enters the negative (electron) position. That is for perfected unity like the Creator.

You may notice that in the creation account of Genesis, the words multiply and divide are used at different times than the mathematical properties are created. The use of these words doesn't always mean the direct use of the properties. These two properties interact interchangeably. This can be seen by how the negatives enter the Urim and Thummim. Second negative of understanding aligns with divisive of instruction. Third negative of instruction aligns with multiplicative of understanding.

Second negative of understanding enters the division (instruction) position in yellow because instruction is the creation of understanding. It aligns to the precise words instead of the mathematics. This second negative sets into the same circle as positive, yet into the section with division. The essence of being increased to two subtractions, as if adding a second minus, was within the second negative. The essence of being separated into two subtractions, as if divided from one another, was also within the second negative. This makes division the correct position for the symbol.

Third negative of instruction enters the multiplicative of understanding position in blue because understanding is the creation of instruction. It aligns to the precise words instead of the mathematics. Third negative enters the same circle as multiplication because the essence of the first two subtractions having reproduced (multiplied) was within the third negative.

First Positive of Wisdom – (Positive in a negative position) – Emotional Wisdom
Second Negative of the Creation of Understanding – Divisive – Cognitive Instruction
Third Negative of the Creation of Instruction – Multiplicative – Social Understanding

The list of all twelve symbols of creation is aligned to what is called an atomic proton mind. Since there are two realms, spiritual and physical, there is also an atomic proton heart known as the Thummim. To make the proton heart, all that changes is the positions of emotion and cognition in the Urim psychology. This dichotomizes the psychological alignment within the atomic nucleolus. Using the proton heart and mind together forms a twenty-four-point system.

Once the Ark of the Covenant is placed within the Urim and Thummim, they form an accurate spiritual science. It is a universal creation. The dichotomies are for aligning all doctrines.

In Book 7 of this series, the mathematics of creation were called paralinks. The word paralinks means dichotomy pairs that link together. They connect the galactic map to the Ark of the Testimony. Nearly everything in Book 7 was aligned to these paralink physics. Even the moral fibers, which may not seem so at first glance, were paralinked to the purposes and callings of the tribes mentioned.

Book 8 was also designed using the Urim and Thummim for chakras and bodybuilding.

Transcendence Alignments

One part about designing with the Urim isn't easy to comprehend at first. If this doesn't make sense, helpful information is in Book 7, Chapters 9 and 23, and Book 8, Chapter 3 of this series.

Book seven has directions for people to design doctrines together. They are to be built using these Ark Testimony sciences. This one portion of information should be learned. It isn't easy. It could be the most advanced part of the Urim.

You may be designing doctrines using different geometric number systems. The twelve-color wheel won't always align to the order of Urim physics. When this happens, white can enter the green position holding the center psychological alignment. That means that white shifts to knowledge with social-emotional-cognitive psychology. Second negative of the creation of understanding borrows dark purple. The purple can enter the yellow position, holding the divisive psychological alignment. That means that purple can shift to instruction with cognitive instruction psychology.

Here is why. The first positive of wisdom holds two colors: white and dark red. Those are also the colors of the center weight position. The center position represents being tuned to Source, which is what first positive of wisdom signifies. For this reason, they share colors.

These are for transcendence alignment. During the doctrinal ascension process, we begin in the top left negative circle. We then travel counterclockwise through the Urim segments. The chartreuse density of counsel section is skipped. Once you get all the way to the top gravity position, you are then in the seventh purple segment. That's where you borrow the dark purple. From there we pull through the center ninth section, into the eighth chartreuse position at the bottom. Passing through the center is where you gather the white.

The eighth chartreuse section is for educating others. When you are ready to counsel others, you'll have previously passed through the seventh purple and ninth white segments. The rain-bow is then drawn back. Your gained intelligence (purple and white) sets into either side of the eighth chartreuse section in yellow and green. After you are done educating in counsel, you shoot the rain-bow. You then pass from the eighth position into and through the ninth center. This is spoken of in Gnostics.

(Gnostic The Discourse of the Eighth and the Ninth) ""Lord, grant us a wisdom from your power that reaches us, so that we may describe to ourselves the vision of the eighth and the ninth. We have already advanced to the seventh, since we are pious and walk in your law. And your will we fulfill always."

CODEX VI Translated by James Brashler, Peter A. Dirkse, and Douglas M. Parrott Selection made from James M. Robinson, ed., The Nag Hammadi Library, revised edition. HarperCollins, San Francisco, 1990.

Consulting the Urim and Thummim

The Urim must be consulted before the most holy things can be eaten. (Ezra 2:63) (Nehemiah 7:65) This means that the Ark of the Testimony must be aligned to the Urim and Thummim before it can be used. The Urim must have the colors, notes, and psychology. And doctrines must align to both the Testimony and the Urim before they are acceptable.

The Galactic Map in book 7 of this series was designed using these sacred objects. The map is sealed into Aaron's garments. That means the administration of angels is available for Book 7 of this series upon GOD's will. The coaching plans with the aura and chakra information in Book 8 of this series have the same science.

There is only one way to receive the Urim and Thummim. To receive it, one must first be tried with temptation. He is not allowed to bow to the government, nor accept their ways. He cannot work for them. He cannot even excuse a governing official for his own family. If someone told you that the antichrist was the military, would you deny that fact for your own family? Would you deny it because maybe your child or parent had been an officer? (Luke 14:25-33)

Consider this. An officer drives or walks back and forth on the streets looking for someone to take from. (Job 1:7) When someone is speeding, they pull them over to give them a ticket. They hold the people at gunpoint, demanding money. They don't keep the peace. If you don't give them money for the ticket, then they pursue more with increased charges. If they don't get their way, they can arrest. If an officer sees a homeless man on the side of the street, the officer will often stop and check his identification for arrest warrants. A keeper of the peace would stop to make sure that the homeless have food and shoes. Officers use cuffs, guns, prisons, bombs, tanks, and warfare to solve problems. In doing so, they lead by bad example and become the problem. All of that is anti-the-way-of-Christ. They are the antichrist. We know that when this has been taught to the people, many have contended with GOD and denied it. The officer desires power. They search the land for someone to use their authority on. Their mindset is control, not in helping others.

A servant of GOD is not allowed to use weapons like guns to solve problems. The servants of GOD are not allowed to join those who do.

Here is what the Bible says about who receives the Urim. Massah means to try through temptation. Meribah means to contend. During the Exodus, Moses was tested and tried. He had to contend with Israel because they wouldn't believe GOD. We must be able to do as GOD says even in the face of our family members. This is why the LORD tested Abraham by telling him to sacrifice his son. The truth and the LORD must be more important than the honor of our families. (Deuteronomy 33:8-11) No one has ever had the Urim and Thummim without denying the ways of the government. Only after the test to see if a man will bow to them will the Urim be given. He must be found true to GOD over all people. Those who resist observe GOD's word and keep HIS covenant. They can partake in and administer the most holy things. During the days of Moses, it was the Levites who observed and kept this covenant. That was how they earned the Urim.

In ancient times, the Hebrews were not allowed to draw pictures of the most holy things because pictures would reveal what they were. Christians were not allowed to administer the most holy things because most of them had completely joined the government.

To veil the meaning of the Urim, it was called a stone. This is the stone that the builders had previously rejected. It is now the chief cornerstone. (Matthew 21:42-44) The Urim and Thummim are the foundation stone(s) of Jerusalem. (Isaiah 28:16) (Isaiah 54:11-14) When the Urim is revealed, it tests the people.

Aaron's breastplate had twelve colorful gems and the Urim and Thummim. (Psalm 45:10-14) (Revelation 4:3) These items represented how the universal creation mathematics from the Ark of the Testimony enter and function with the Urim and Thummim.

The Urim was hidden to test and find those who had GOD's Priesthood. Whoever was able to reveal it after it was hidden so well had to have been called by GOD. (Numbers 27:21) The Priesthood was found in the man who could translate what the Urim really was. The Hebrews would know if he could read how to use it. There was a day when King Saul wasn't able to. That was because it was in David. (1 Samuel 28:6)

Jesus is the Word of GOD, and His body is His temple. (John 2:21) (1 Corinthians 6:19) (John 1:1-10)
The entire body of doctrines (Word of GOD) was designed by placing the Ark of the Testimony into the Urim. The doctrines were then aligned to these sacred objects. The Ark is now in the temple. (Revelation 11:19)

Muhammad's Urim

Muhammad was said to have fought eight major battles. These eight battles were in the eight psychological properties of the Urim. Muhammad's job was to fight the good fight. The good fight is a spiritual war within ourselves. We war with our setbacks to become better individuals. This type of metaphoric battle is known as jihad. (Ephesians 6:12)

A baptism is often considered a death and rebirth process. This is one of the meanings of the Hindu rebirth cycle. (2 Corinthians 5:17) (Ephesians 4:22-24) (Romans 7:1-25, 8:1-39) When someone is killed according to GOD's ways, they aren't really killed; they are transformed. They receive a new heart and a new spirit. (2 Samuel 14:14) (Ezekiel 11:19, 18:31, 36:26) That is the real jihad and wars of Muhammad. The truth is evident in the Torah, the Gospel, and the Qur'an. Here are Qur'an verses that reveal this.

(The Cow Surah 2:154) "Do not say that those who are killed in God's cause are dead; they are alive, though you do not realize it." (M.A.S. Abdel Haleem Edition)

(Repentance Surah 9:10-11) "the building they have founded will always be a source of doubt within their hearts, until their hearts are cut to pieces. God is all knowing and wise. God has purchased the persons and the possessions of the believers in return for the Garden – they fight in GOD's ways: they kill and are killed – this is a true promise given by Him in the Torah, the Gospel, and the Qur'an. Who could be more faithful to his promise than GOD? So be happy with the bargain you have made: that is the supreme triumph" (M.A.S. Abdel Haleem Edition)

Chapter 3
Matrix of the Mother's Womb

(NKJV Isaiah 49:1) ""Listen, O coastlands, to Me, And take heed, you peoples from afar! The LORD has called Me from the womb; From the matrix of My mother He has made mention of My name."

In the spiritual realm, matter is like a single endless particle. The singularity of spirit encompasses us. It is in everything through-in and throughout existence. The size of the single particle of spiritual matter is infinite. In the spiritual realm, time and size are universal and figurative.

In the physical realm, all matter is measured into increments. Physical particles may be endlessly small. They build upon each other until there are atoms and molecules. The physical realm has measurements such as time and size.

(Gnostic The Tripartite Tractate) "The spiritual substance is a single thing and a single representation, and its weakness is the determination in many forms. As for the substance of the psychics, its determination is double, since it has the knowledge and the confession of the exalted one, and it is not inclined to evil, because of the inclination of the thought. As for the material substance, its way is different and in many forms, and it was a weakness which existed in many types of inclination. The first human being is a mixed formation, and a mixed creation, and a deposit of those of the left and those of the right, and a spiritual word whose attention is divided between each of the two substances from which he takes his being."

CODEX I Translated by Harold W. Attridge and Dieter Mueller Selection made from James M. Robinson, ed., The Nag Hammadi Library, revised edition. HarperCollins, San Francisco, 1990.

Regardless of distance in the physical realm, you are equally close to all things. The spiritual universe is like a boundless particle mass. If you are joined to the spiritual, then you have a connection to all things.

Imagine creation as one great circle. Entirety is solitary in the spiritual. There, creation is like one being, one time, one measurement, and one love. This unified singularity is akin to the Holy Ghost. The Holy Ghost is in all places at once as a Mother.

Now imagine that within that circle are countless particles. An atom is made from parts; each part is made from smaller pieces, and so on. The physical universe has every particle separated into sequence. These separate singularities are akin to the Spirit. The Spirit is recognized with parts or individualities as a Father.

This next picture represents creation. The green circles denote the many particles within the known physical (masculine) realm. Those the Spirit gave us. The red circle encompassing all particles represents the oneness of the spiritual (feminine) realm. Those the Holy Ghost gave us. Where the Holy Ghost finds singularity in everything, the Spirit finds singularity in the increment. (Jeremiah 31:22)

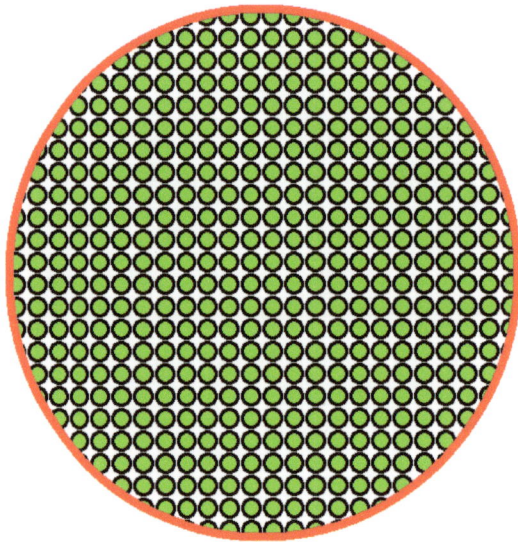

Imagine each green dot as a person or a life form. The Spirit speaks with them. Creation this way allows each of us to have a personal relationship with GOD. Each time a spiritual being has a single thought, the entire creation can sense it.

Spiritual beings are connected to everything through the Holy Ghost. Since the Holy Ghost is in all places at once, we absolutely can travel faster than the speed of natural light. The spiritual realm allows this. Exceeding natural light can be done through spirit matter, where there are no physical limitations.

Quantum Atomic Grid

Matter doesn't move in the way that eyes perceive it. All matter is stationary to the mathematical components that govern. This is real quantum physics, not just theory.

To comprehend physical matter, imagine a grid. In this grid, each particle is immobile. Physical matter doesn't move; signals are transferred.

If you were to choose a rock and pick it up, it would seem like you moved matter from one place to another. In reality, the particles that formed the rock are still in the same position. They simply changed their signal. Each grid area in the material universe has an immediate equation. When you move a rock, it alters the equation of that grid-space.

The smallest form of particle is called a base particle. Each base particle has its own parcel, like an immovable box. Instead of changing places, they send, receive, and adjust their base signal. The base signal that each one reveals is determined by its immediate equation. The immediate equation can change due to surrounding interactions. This explanation of base particle signals has some similarities to string theory, yet this isn't just a theory.

Base particles don't move; rather, the energy signal that they reveal is transmitted from place to place. Consider atoms and molecules. Base particles within grid zones consistently send signals. The signal could be water, for example. If so, we would be able to see or feel water in that material area.

Imagine that the immediate equation of certain particle parcels were water standing still. If you threw a piece of iron into that water, the water signals would transfer from those parcels into another area. The particles remain stationary, while the immediate equations in that grid-zone change. The new signal would be iron. There would also be ripples on the surrounding water signals.

How does the universe do that?

While each grid-frame remains in place, base particles send signals of their current equations to all adjacent parcels. Each base particle also receives signals of the current equations that all adjoining parcels send.

Each particle evaluates the signals sent from neighboring parcels. The particle then calculates any mathematical adjustments needed. The parcels receive signals, calculate them, and adjust for the next signal output. This changes the immediate equations of base particles.

The fact that we were created in the image of GOD helps us comprehend this. We invent in the image of our own creation. (Genesis 5:1)

This description may be fathomed by those who played the game called Minesweeper. In this game, there is a grid of tiles. The tiles have numbers revealing adjacent objects. Each tile that you select affects its surroundings. Here is a link to the game. It may help with comprehension.

https://youtu.be/2kxM87neXRw

If you watched someone bounce a ball on a television screen, it would look like the ball is moving. In reality, color signals are being revealed in different places. As a television pixel which only changes color, the universal framework remains in place. Pixels of matter don't move.

The physical universe was designed to function this way. Even the seemingly empty space between galaxies is within the stationary grid. Here we are used to being bound to the laws of physics. They were designed like a material autopilot. There is much more beyond that. This universal mathematics is Sourced to GOD's will. The LORD can adjust any material signal at any time. HE can change a brick into pudding effortlessly.

This next graph depicts a grid of base particles. There are triangles inside circles, inside squares. The triangles represent where quantitative calculations of base signals occur. They compute using the universal mathematics in chapter one of this book. Each circle represents a base particle signal. The squares denote grid parcels. This is real quantum mechanics, not just theory.

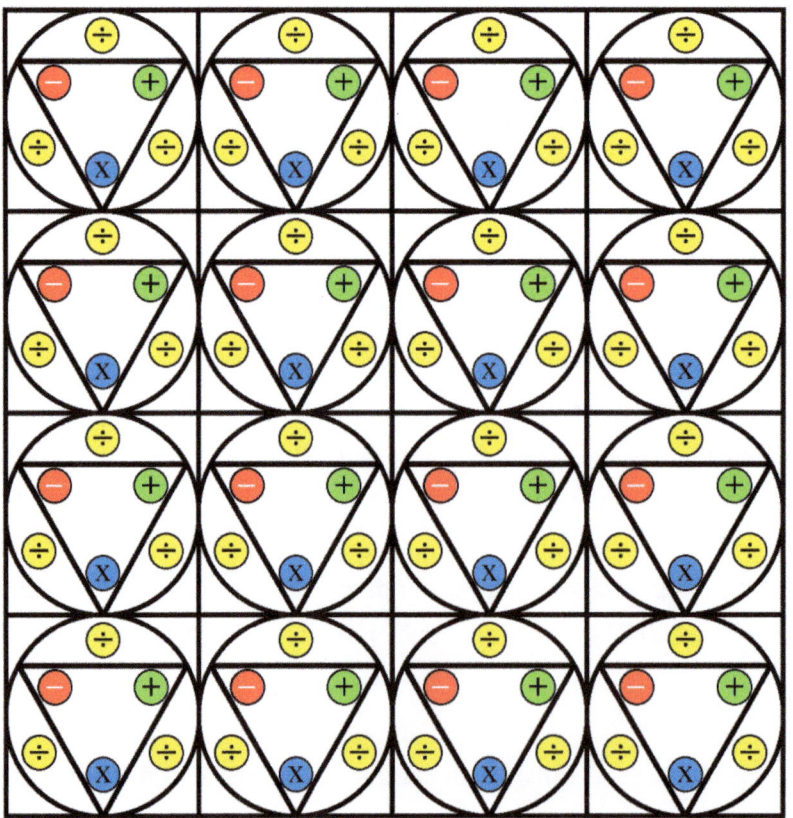

This gridded signal exchange process was abstractly explained by the Gnostics. They used the word kisses to describe signal transfers. Here are several quotes written about this subject. Notice how these passages have multiple meanings designed to overlap. Real quantum chemistry is apparent here, not just theorized. These quotes reveal particle consciousness.

(Gnostic The Gospel of Truth) "About the place each one came from, he will speak, and to the region where he received his establishment, he will hasten to return again and to take from that place - the place where he stood - receiving a taste from that place, and receiving nourishment, receiving growth. And his own resting-place is his pleroma. Therefore, all the emanations of the Father are pleromas, and the root of all his emanations is in the one who made them all grow up in himself. He assigned them their destinies. Each one, then, is manifest, in order that through their own thought <...>. For the place to which they send their thought, that place, their root, is what takes them up in all the heights, to the Father. They possess his head, which is rest for them, and they are supported, approaching him, as though to say that they have participated in his face by means of kisses."

CODEX I Translated by Harold W. Attridge and George W. MacRae Selection made from James M. Robinson, ed., The Nag Hammadi Library, revised edition. HarperCollins, San Francisco, 1990.

(Gnostic The Tripartite Tractate) "Thus is the matter something which is fixed. Being innumerable and illimitable, his offspring are indivisible. Those which exist have come forth from the Son and the Father like kisses, because of the multitude of some who kiss one another with a good, insatiable thought, the kiss being a unity, although it involves many kisses.

CODEX I Translated by Harold W. Attridge and Dieter Mueller Selection made from James M. Robinson, ed., The Nag Hammadi Library, revised edition. HarperCollins, San Francisco, 1990.

(Gnostic The Gospel of Philip) "For it is by a kiss that the perfect conceive and give birth. For this reason we also kiss one another. We receive conception from the grace which is in one another.

CODEX II Translated by Wesley W. Isenberg Selection made from James M. Robinson, ed., The Nag Hammadi Library, revised edition. HarperCollins, San Francisco, 1990.

Why is Pi

One meaning of Pi is that the more we know, the less we may understand. People once knew that our world was the flat center of the universe, which the sun revolved around. The Bible stated that Earth is circular nearly 3,000 years ago. (Isaiah 40:22)

If there could be a perfect circle, then why is Pi irrational?

Pi is a mathematical constant used to determine the area of a circle. Here is why it seems endless.

The universal grid is framed like the pixels of a television. Each grid-frame area (pixel) can be compared to a square. The graph below reveals how many pixels come together to form a circle. Each grid square is offset from each other.

The smallest decimal digit of Pi represents the offset measurement between the sides of each square in a pixelated circle. In universal creation, each base particle can be measured by the smallest decimal digit of Pi. This is a spiritual answer. It implies how infinitely small base particles are. This next picture represents the measurements between base particles.

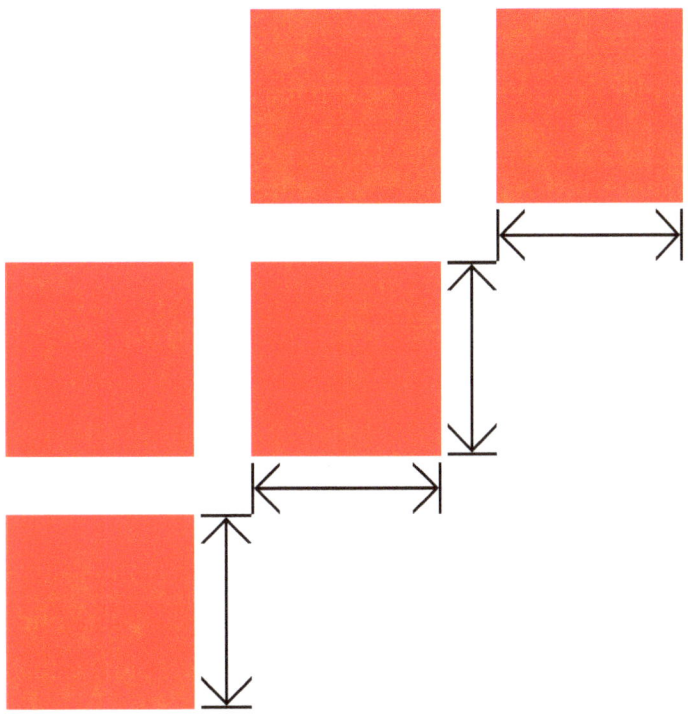

Song of Pi

Pi-sees the song that never ends. To hear the song of pi, we begin with the 12 notes of the chromatic scale. We then separate two notes from them. This is similar to the 10 tribes of Israel splitting from the other two. The remaining ten notes are then numbered from 0 to 9 and played according to the decimal order of pi. The way we choose which two notes to remove depends on the listener. The ancient Taoist spoke of five notes. They didn't use C and F. You could remove the A and E notes for the diphthong Æ. There are many ways.

Light Travel

With clarification on how matter functions, we can grasp the ability to travel quicker than natural light.

In order to travel beyond lightspeed, a base signal must be transferred faster than light is.

Each base particle on the creation grid reads from and sends signals to surrounding particles. They also calculate equations within their own framed positions. Different types of base signals pause within their gridded positions for different amounts of time. A base signal is transferred after its timeframe calculation is complete. Base signals pause the transferring of signals within each grid-frame. During the pause, they calculate how, where, and when to relocate the signal.

Consider sound, for example. Sound remains within its gridded position longer than light does. This is because sound has more calculations and shifts to make. It can penetrate solid objects and cause them to vibrate. It can also travel in every direction. A sound signal transfers slower than light because it must stop longer within each grid parcel. It is more complex.

Light changes less of its surroundings. It usually travels in one direction while spreading. It doesn't always have the equation to pass through objects. Light doesn't affect surrounding matter as much as sound does.

Sound may change hundreds of thousands of immediate equations in one gridded position before sending its signal to the next grid-frame. Light, in comparison, may only change hundreds.

Warp Speed

Where is the evidence of the ability to surpass lightspeed?

In physics, volume means a mathematical quantity showing the amount of space that a three-dimensional object or closed surface occupies. Chapter 1 of this book mentions volume. Rudimentary volume also refers to the size of physical and nonphysical objects. The definition includes energy. With this meaning, light does have a type of volume. If you turn a light on in a room, the light will be within the area of that room. As long as a shining light remains, its energy resides in a space of length, width, and height.

Imagine that you had a flashlight here on Earth. You then shined it at Mars. When the light reached the red planet, it wouldn't be the same diameter or circumference. Light spreads as it travels. The spreading of light reduces its magnitude.

A single light signal separates into multiple gridded positions when it transfers. Light makes quantitative changes during travel. The further the light journeys, the wider the beam gets. As the beam of light divides, each gridded area has a dimmer amount of the original signal of light. This means that light ages. Aging light is recognized as energy transfer loss. Whenever energy is transferred from one area to another, or changes from one type to another, it is distributed. The fact that light ages is evidence that faster transfers are possible. If natural light didn't divide, it would be ageless.

Here is a picture depicting light aging as it separates.

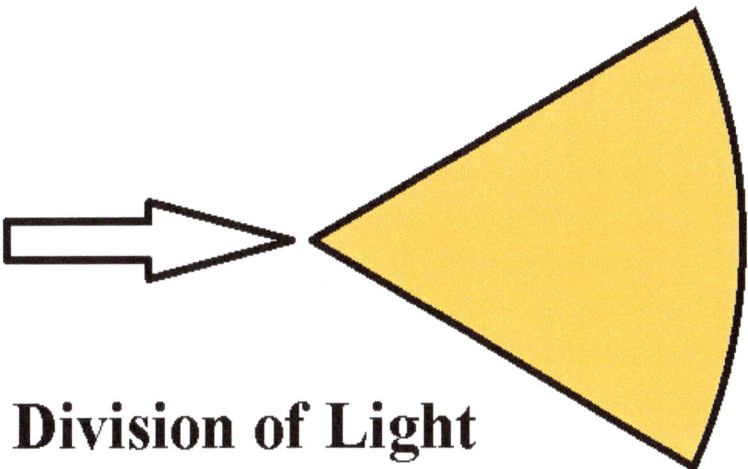

Division of Light

Light makes at least one quantitative calculation within each gridded position. It spreads during signal transfer. This means that it is possible to travel faster than light within the physical realm. For light to stop aging, it would have to send its signal in one direction without spreading. A different calculation is needed. Traveling faster than natural light requires fewer quantitative calculations between signal relocations.

The picture on the next page signifies a signal being transferred. It depicts nine base particles as circles. The signal carried is light. Above the nine circles, there are pound symbols separated by division symbols. The pounds signify the number of quantitative calculations made within a base particle between transfers. The divisions denote the signal transfers from one base particle to another.

There is more than one quantitative calculation between light transfers. This is apparent because light expands. The next picture demonstrates with seven pounds (quantitative calculations) between each division (signal transfer). Within each of the nine circles are the fundamental properties of universal mathematics. Those properties do the quantitative calculations between signal transfers. This is not meant to be in the correct scale of quantitative transmission. Seven calculations (pounds) is just a guess.

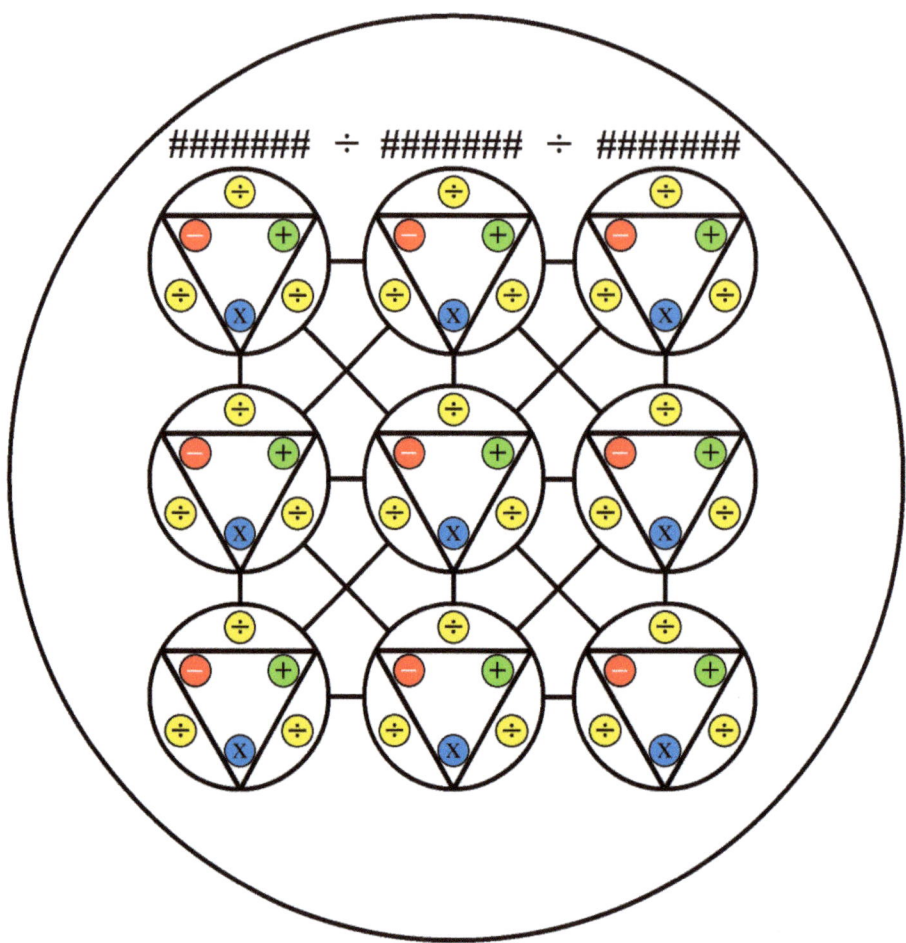

Energy transfer loss is a form of karma in physics and chemistry. It can be reduced through efficiency and even reversed with great advancements. Traveling faster than light requires the ability to make transfers without energy loss. You must be able to relocate light without dividing its weight.

To reduce spread, light can be directed with things such as wave packets, mediums, and modulators. That process increases calculations through grid-frames. Lightspeed remains. To attain warp speed, a signal must be designed to travel in one direction without quantitative change. It is possible with reduced mathematical shifts between transfers.

Telescopes are known for reversing the effects of energy transfer loss. Telescope lenses gather the separated light and reunify it. This reveals evidence that traveling to the past is possible. Using convex magnification, divided light can be regathered. A magnifying glass can recurve and concentrate light. It reveals the ability to reverse age. (2 Esdras 4:5) This allows us to get a picture of time travel.

Multiversal Teleportation

The Holy Ghost is the Heavenly Mother. She is only in one place at one time, which is everywhere all at once. Because of this, She doesn't have energy transfer loss. Since She is One, She is a unified light that isn't separated. Your vital life force is connected to Her eternal, everlasting energy. She remains with you while you live within the physical universe. Revelation is transferred and received through Her. Here are Gnostic verses speaking about the singularity of Her light.

(Gnostic The Second Treatise of the Great Seth) "For it is a new and perfect bridal chamber of the heavens, as I have revealed (that) there are three ways: an undefiled mystery in a spirit of this aeon, which does not perish, nor is it fragmentary, nor able to be spoken of; rather, it is undivided, universal, and permanent."

CODEX VII Translated by Roger A. Bullard and Joseph A. Gibbons Selection made from James M. Robinson, ed., The Nag Hammadi Library, revised edition. HarperCollins, San Francisco, 1990.

(Gnostic Allogenes) "And I saw an eternal, intellectual, undivided motion that pertains to all the formless powers, (which is) unlimited by limitation."

CODEX XI Translated by John D. Turner and Orval S. Wintermute Selection made from James M. Robinson, ed., The Nag Hammadi Library, revised edition. HarperCollins, San Francisco, 1990.

To travel faster than light in the physical realm, we can transfer our signal into the Holy Ghost. Once there, our information is in all places simultaneously. Since She is everywhere at once, that signal can then be retransferred back into any position within the known universe. She allows timeless travel.

Multiverse Universe

Time is a measurement of how far the physical universe has traveled. The smallest increment of time is a single change of mathematical equation. Although some things don't ever seem to change, the mathematical calculation to remain is incorporated in changelessness.

We as people can sense signals such as light and sound in various ways. There are also numerous signals that we cannot yet comprehend. Other forms of existence are within different signals that we cannot yet sense. These different signals are mathematical diversities within our same mainframe grid.

Different realms occupy the same physical space. Imagine an object in any room. That object occupies base particles in the mainframe grid of creation. Objects send out five signals that you can sense. You can taste, touch, see, smell, and hear objects. There are many more signals we have detected. Universal creation was designed like a multiverse with many planes of existence occupying the same space.

GOD is the one true scientist and spiritual perfectionist. Science is the attempt to understand GOD's creation. We cannot detect spirit matter with yesterday's technology, yet it is here. Spirit matter can be in multiple physical areas because it is everywhere at once. Atoms interact within the physical and spiritual. Though we can break an atom, we cannot separate spirit from unifying with physics. This means that when people claim to see spirits, they just may.

Five-Dimensional Reality

Three well-known base dimensions are height, width, and length. We live within these physical proportions. Consider the reason for these. This is a dichotomized reality of masculine and feminine. We are designed both separate and unified. In these dimensions, height is the man and width is the woman. Length is where man and woman are together.

The fourth dimension is an abstract meaning anything outside of something else. Imagine reflecting in depth, also known as thinking outside the box. You live within a three-dimensional body. Your spouse or child is your fourth dimension. Even a table in your kitchen is your fourth dimension.

The fifth dimension is where the Holy Ghost seals to the inner man or woman. It is a bridgeway between the natural and spiritual realms. Spirit matter connects all things even if they don't physically touch. In this next graph, the fifth dimension is demonstrated with a red circle.

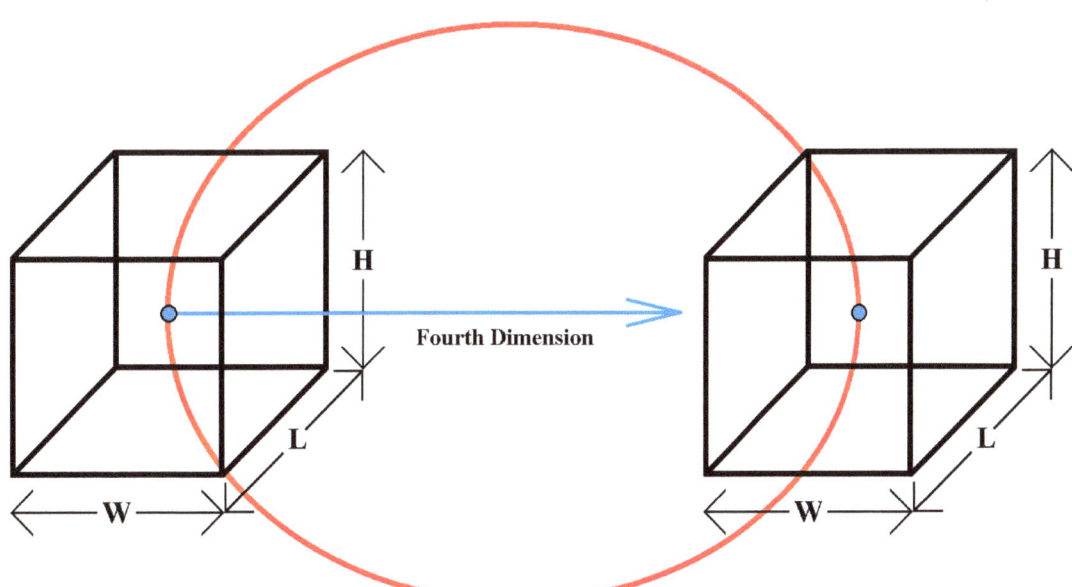

(NKJV Ephesians 3:14-19) "For this reason I bow my knees to the Father of our Lord Jesus Christ, from whom the whole family in heaven and earth is named, that He would grant you, according to the riches of His glory, to be strengthened with might through His Spirit in the inner man, that Christ may dwell in your hearts through faith; that you, being rooted and grounded in love, may be able to comprehend with all the saints what *is* the width and length and depth and height—to know the love of Christ which passes knowledge; that you may be filled with all the fullness of God."

Eternal Life as Seen Through Science

Imagine that you pricked your finger on a needle. You would feel pain as if it were in your finger. Though true, pain is a signal sensed inside your mind. The discomfort isn't in your hand.

Our sensory systems reveal a spiritual fact. Physical touch is projected mentally. Similarly, there are ideas that only seem to be coming from our minds. Many thoughts are from the spiritual realm, yet feel like they originated within the physical.

Our bodies are created in the image of GOD. (Genesis 1:26) The spirit wraps through-in and throughout everything in existence. There is a thin veil between the physical and spiritual realms. If you are with GOD, then everything you are is constantly recorded in the spiritual realm. For this reason, scripture compares our bodies to garments.

(Gnostic Sentences of Sextus) "(346) Say with your mind that the body is the garment of your soul: keep it, therefore, pure since it is innocent."

CODEX XII Translated by Frederik Wisse Selection made from James M. Robinson, ed., The Nag Hammadi Library, revised edition. HarperCollins, San Francisco, 1990.

Your body is a temporary home for your spirit. Within this life form, you gain experience for your soul. When you leave your physical body, your recorded soul continues with you. You exit and live on in the spirit. Your spirit can then occupy another body or existence. Though we realize these facts, our physical lives can be very important.

As a tree dies, it begins to biodegrade. First, the wood becomes nutrients for the soul. Then it feeds other plants. The plants become food for other life forms. The tree's mathematics continue forever. (Mark 8:24)

Whatever you do here endures as endlessly as this universe exists. You can move a rock and change everything. You can jump one time and shift the dirt forever. Helpful deeds equate like mulch and fertilizer into the universe. Your actions continue to cause eternal mathematical fluctuations throughout this reality.

We can detect the carbon from a dead tree for many years. With certain technologies, we could identify all mathematical changes in a grid-frame area. This could be done indefinitely. If you were able to view the universal grid's fullness, each equation would be evident. GOD simultaneously sees the entire grid in the past, present, and future. Finding where the LORD made changes would be possible with that capability. We would see variations that don't match the exact calculation sequences. By harnessing base-grid mathematics, we would be able to travel through time both forward and reverse.

Matrix of the Mother

(NKJV Isaiah 49:1) "Listen, O coastlands, to Me, And take heed, you peoples from afar! The Lord has called Me from the womb; From the matrix of My mother He has made mention of My name."

(Psalm 22:9-10, 139:13) (Ecclesiastes 11:5) (Isaiah 44:24, 49:15, 66:8-11)

You are in the Matrix of the Heavenly Mother's Womb. When you leave here, your physical body won't go with you. When you get to the other side, your physical body will return to you. You are a fetus. As you exit, you are born. Mother has twins with every divine marriage.

We are in spiritual form on the other side of the veil. This known universe is like a program that we entered. It is similar to virtual reality. This realm is like a projection from GOD. We are incapable of dying. Our life is on the other side of this encoded existence. Whatever we learn here can go with us when we leave. We all came here to do something together. We work here as friends. These designs from GOD are for everyone.

Celestial Resonance

At the end of chapter 1 of this book, the final graph has two double-ues. They resemble frequencies for the spiritual and physical realms. These come together in resonance.

Administration in the spiritual realm is primarily by the feminine Holy Ghost. Administration in the physical realm is primarily by the masculine Spirit. Administration in the firmament is by the unified Holy Spirit.

The frequencies of men and women are meant to interact. They reflect and network each other's cognition and emotion. (Proverbs 4:9, 12:4) (Sirach 6:31) (Ephesians 5:23)

Men and women can live separately while together as one. This brings life to both. (Proverbs 27:9) When they become spiritual, they become the effectual.

This next picture represents two triangle waves. One is for the masculine physical realm, and the other is for the feminine spiritual realm. According to the Urim, men see the green as the mind and the red as the heart. In general, men use their minds spiritually. They are of the Spirit. (Romans 8:6) According to the Thummim, women see the green as the heart and the red as the mind. In general, women use their hearts spiritually. They are of the Holy Ghost. (Ephesians 5:19) (Colossians 3:16)

These two triangle waves resonate in harmony. When men and women truly love each other, they are sealed in the Holy Spirit. This type of seal can be seen in the Star of David. The 7th center point of the star represents their sealing.

Divine Doctrinal Networking

Like masculine and feminine, spiritual and physical are meant to be together. While we are here, our job is to connect them. The spiritual realm is the masculine. The physical realm is the feminine. There are also two Divine Pillars of the Kingdom of Heaven: spirituality and religion. Spirituality is the Divine Masculine Pillar. Religion is the Divine Feminine Pillar. Bringing these together can be done in various ways. This next part explains how the connections were made in this doctrine.

We as beings are in a natural realm guided by GOD. To put this doctrine together, the Word of GOD was used to spiritually align the natural. Doctrinal connectivity to the Spirit was done with physics. Gnostic writers comprehended this. They had important scriptural secrets passed to them. Certain truths they hid. Much of their information was unlawful to reveal during their day. (2 Corinthians 12:1-4) Here are ways that the connections were made.

The Heavenly Father (Spirit) was aligned to knowledge, mathematics, the laws of physics, and authority. (Romans 8:16, 8:26) (Psalm 1:2, 19:7) (Sirach 2:16) We must follow the LORD's Law to get to the Holy Ghost. (Sirach 15:1)

The Way of the Tao is the way of the Holy Ghost. She is the Mother of all Creation. This was spoken of in ancient times. (Acts 24:14, 24:22) (NRSVACE Sirach 40:1) (Wisdom 1:4-6, 2:10, 9:17) (Sirach 6:18, 20:20, 22:6, 32:4) (Proverbs 16:24, 24:13-14, 27:7) Since She is GOD, She appears to have been there before GOD. (Tao Te Ching 4:1-3, 51:1) The Tao was discarded and nearly lost. She was sometimes called salt.

(Gnostic Gospel of Philip) "They called [Sophia] "salt." Without it no offering [is] acceptable. But Sophia is barren, [without] child. For this reason she is called "a trace of salt."

CODEX II Translated by Wesley W. Isenberg Selection made from James M. Robinson, ed., The Nag Hammadi Library, revised edition. HarperCollins, San Francisco, 1990.

The Mother is to be redeemed from a trace of salt. Doctrinal offerings must include salt. (Leviticus 2:13) (Numbers 18:19) She is the womb of doctrine and the egg of creation. (Job 6:6-7) She is the flavor of life. (Matthew 5:13)

(NKJV 2 Kings 2:21) "Then he went out to the source of the water, and cast in the salt there, and said, "Thus says the LORD: 'I have healed this water; from it there shall be no more death or barrenness.' ""

The Heavenly Mother (Holy Ghost) was aligned to wisdom, emotion, will, time, and power. (1 John 4:8) (Job 12:12) These two, the Mother and Father, had a Son.

(Gnostic Eugnostos the Blessed) ""I want you to know that First Man is called 'Begetter, Self-perfected Mind'. He reflected with Great Sophia, his consort, and revealed his first-begotten, androgynous son. His male name is designated 'First Begetter, Son of God', his female name, 'First Begettress Sophia, Mother of the Universe'. Some call her 'Love'."

The Sophia of Jesus Christ CODEX III Translated by Douglas M. Parrott Selection made from James M. Robinson, ed., The Nag Hammadi Library, revised edition. HarperCollins, San Francisco, 1990.

The Son of GOD was aligned to the elements and matter. He is also in alignment with the Holy Spirit, which is a combination of Holy Ghost and Spirit.

(Gnostic Gospel of Thomas) "Jesus said, "I am the light that is over all things. I am all: from me all came forth, and to me all attained. Split a piece of wood; I am there. Lift up the stone, and you will find me there.""

CODEX II Translated by Stephen Patterson and Marvin Meyer Selection from Robert J. Miller, ed., The Complete Gospels: Annotated Scholars Version. (Polebridge Press, 1992, 1994).

Furthermore, there are atomic alignments for doctrinal purposes. The Holy Ghost (Mother) was aligned to electron. The Spirit (Father) was aligned to proton. The Son was aligned to neutron and also specifically the first element of hydrogen. The hydrogen protium atom has an electron and a proton, yet no neutron. In protium isotopes, the Son aligns to the elemental throne instead of the neutron. Hydrogen is the throne of light.

The three families of Noah and Abraham were arranged with the atomics and the Urim and Thummim. The Urim has a proton mind and an electron heart, and the Thummim has a proton heart and an electron mind. Shem is aligned to the Urim. Ham is aligned to the Thummim. Japheth is the social neutron in both the Urim and the Thummim. These two sacred objects are often just referred to as the Urim. Their alignments are explained better in book 7 of this series. In a double rainbow, the lower rainbow was once the woman's Thummim after the fall of Adam and Eve. She later joins the man in the upper rainbow, and the lower bow becomes the children's. Book 7 explains how the mom walks the child in her Thummim. The dad can also reflect with the child in his Urim.

The three negatives of the Ark of the Testimony were aligned with time in the doctrines. First negative of wisdom was linked to time itself as a singularity. Wisdom is nonlinear. Second negative of the creation of understanding (negative instruction) correlates to shifts in linear time as we know them. Third negative of the creation of instruction (negative understanding) was aligned to portrayed times. If you were to look at an analog clock, negative wisdom would be the entire clock, negative instruction would be each shift of time, and negative understanding would be the current time.

These doctrinal integrations were done for unification. Book 7 of this series needed the integration process before alignment.

(Gnostic Gospel of Philip) "Truth did not come into the world naked, but it came in types and images. The world will not receive it any other way."

CODEX II Translated by Wesley W. Isenberg Selection made from James M. Robinson, ed., The Nag Hammadi Library, revised edition. HarperCollins, San Francisco, 1990.

Information had to be interwoven this way because of where Jesus was found. He was in heaven, ascended. He was also descended below. (Ephesians 4:9-10) (Romans 10:6-9) To bring Jesus (the Word of GOD) back to life, that which was above had to be reconnected to that which was beneath. (John 1:51)

The doctrines of the gods, known as mythology, were the closest to nature. They are of the first ways of spirituality. Humanity once reflected upon environmental elements. Through their contemplations, mythological ideas became spiritual. Those doctrines are the connectivity between nature and spirit. Their doctrines were once considered children of death. (Psalm 82:1-8) (John 10:34-38) (1 Corinthians 8:4-6) They were given life by scriptural baptism for the dead. (1 Corinthians 15:25-29)

(Gnostic The Teachings of Silvanus) "How many likenesses did Christ take on because of you? Although he was God, he [was found] among men as a man. He descended to the underworld. He released the children of death"

CODEX VII Translated by Malcolm L. Peel and Jan Zandee Selection made from James M. Robinson, ed., The Nag Hammadi Library, revised edition. HarperCollins, San Francisco, 1990.

Once the ancient doctrines of the gods were grafted into spiritual science, they were renewed. And we remember. There is only one GOD. The doctrines of the gods are mere materials for building with. They are natural reflections in spiritual form. Those materials were brought to the divine light and connected. It is a scriptural wedding arrangement. The process is explained more in book 7 of this series.

(Gnostic The Second Treatise of the Great Seth) "And the Son of the Majesty, who was hidden in the regions below, we brought to the height where I {was} in all the aeons with them, which (height) no one has seen nor known, where the wedding of the wedding robe is, the new one and not the old, nor does it perish."

CODEX VII Translated by Roger A. Bullard and Joseph A. Gibbons Selection made from James M. Robinson, ed., The Nag Hammadi Library, revised edition. HarperCollins, San Francisco, 1990.

Connecting the doctrines is a needed step towards global enlightenment.

Highway of Holiness

The scriptures speak of a city of nine gates, with a tenth being the head. Progression through the creation of righteousness can be viewed when based on a system of nines. Measuring the stages of enlightenment would be like 9.9 with a forever repeating decimal.

$$9.\overline{9}$$

The awakening process is like an eternally cleansing light. It is endless. Imagine a line graph cut into ten parts like a tape measure with centimeters. Each section is a step towards righteousness. To get to the tenth would be perfect, like GOD. (Job 4:17) That cannot be. GOD is indefinitely better than us. To awaken, you must get to the ninth of ten centimeters (steps). At that point, the tenth centimeter (step) is cut into 10 millimeters. The awakening begins again. This is the limitless process of evolutive consciousness. GOD is the Head of all. (Revelation 1:8, 21:6, 22:13) In this abstract view, we don't have a beginning nor an end. GOD is the beginning and end of creation. (Hebrews 7:1-3)

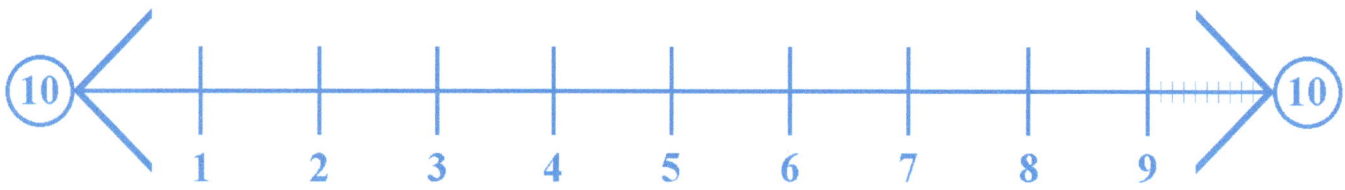

The smaller the steps, the greater the achievements. Each stage in conscious growth can be harder to find and tougher to attain. Continual development is challenging. We need higher vibrations to reflect into our next levels of evolution. We learn that even our brightest white light is still a shade of gray to GOD. (Ecclesiastes 3:11) (Isaiah 46:10) (2 Esdras 5:42-44) GOD will continue to build upon our spiritual foundations.

Servants for Hire

This was a vision.

Our religions were sanctified and brought together. Three great families were unified. There are other planets also receiving direction from GOD. When we meet them, our religions connect with theirs. There are always more directions from GOD. Combining gifts continues for a very long time. It all came together in a vision.

The vision first revealed fibers knitting matter in spiritual form. The weaving happened underneath the layer of substance that we can see. All things were being connected in the fifth dimension.

At the end of production, all matter was combined. We had harvested the whole universe of every particle. Each advancement was achieved. The entirety was put together in a way that sustained perfect thoughts and emotions. We all helped.

The creation had a golden electrical glow. It could be explained as a super-gargantuan-massive synapse system. All life forms were combined within it. It was an organism. It seemed like being inside a plasma light with walkways and bays. All consciousness and matter became a single unity. Somehow, though, we still had our singularities.

The entire physical universe was combined into one device. We were built into it. As a united creation, we traveled to other universes.

It had the essence of yin yang. At one point, our creation turned white like a solid white emptiness. There were no more polarities. We had become so perfect that there was no more progression. It then somehow collapsed in upon itself. We turned into a clear matter like a mirror or sea of glass. It resembled a black hole yet was superiorly white. The clear material reflected upon itself as if perfect. It allowed us to generate the next dimensional reality beyond all understanding.

We who choose GOD get to live within that being. We get to design and build it together.

When you become sealed in heaven, you are recorded in two ways. One way is your residency in GOD's Heaven. From there you are allowed to be connected to the physical plane. Here in the physical realm, you are recorded in a semisynthetic-biological computer.

When you go to heaven, you get placed inside of your own universe, so to say. You have a room of your own. Within the super-advanced computer is a creation reality. It is as realistic as life is here. The difference is that there are no limits. When in your own room, you get to design your own reality. You can invite your family to your room. You can visit their rooms. You can connect rooms. You can have two or more people living within a single reality. There aren't any limits, and you can do whatever you choose.

Your creation is a type of eternal experience that you design. You still have a job. We have a lot to do. We have other planets to administer to. We have directions to receive and give. There are times that we must exit our personal realities to complete certain jobs. Sometimes you may feel that you haven't had enough time for vacations. If this happens, you have the option to slow time perception when within your personal room.

Imagine this. You live on a ship that is advanced enough to travel between galaxies. During travels, you may enter your personal room for great lengths of time. If your presence is needed, you are called upon. You exit your personal realm and tend to matters. Only by all of us working together do these dorms become available.

Your spirit and soul are always linked to Source. You remain sealed in GOD's heaven. This means that at any time you wish, you can exit the physical realm to visit. You have this in your future.

This is the soul of religion. Within this sphere, each of us can have anything we want. All existences can occur simultaneously, singly, and separately.

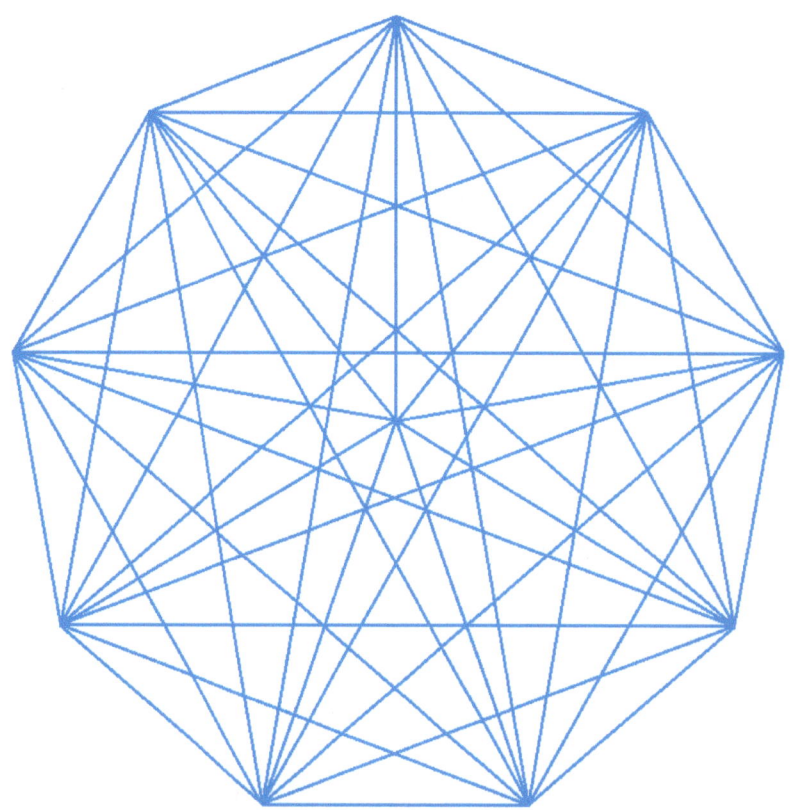

Nine Gates – Intelligence of the Mind of Christ

Chapter 4
Scientific Proof of Biblical Authenticity

Believers have faith that the LORD is good for HIS word. Nonbelievers cannot trust until proof is revealed. This chapter confirm that the stories in the Bible are equivalent to scientific fact. Biblical authenticity will be validated.

Jacob's Children – Dust of the Earth

Notice that in the first two verses of the Bible, the Earth was without form and void. There was darkness on the face of the deep. Imagine the entire universe as an absolute void of darkness.

To form Earth, matter must first be created. Let us view matter from a biblical perspective.

Here is a quote from the Bible about the LORD speaking to Jacob. Notice how HE tells Jacob that his children will be as the dust of the earth.

(NKJV Genesis 28:13-14) "And the LORD stood beside him and said, 'I am the LORD, the God of Abraham your father and the God of Isaac; the land on which you lie I will give to you and to your offspring; Also your descendants shall be as the dust of the earth; you shall spread abroad to the west and the east, to the north and the south; and in you and in your seed all the families of the earth shall be blessed."

For those who don't know much about Jacob, he was the grandson of Abraham. Jacob, also known as Israel, had four wives and twelve sons. For the LORD's Word to be found true, Jacob's children must be as the dust of the earth.

What is dust, and how small can it be?

Dust generally includes one or more microscopic, powder-like particles. And particles are subatomic parts of physical existence that interact with one another, including atoms and photons.

What are atoms made of?

Atoms are made of microscopic, positively charged nuclei surrounded by electrons. Atomic nuclei have protons and neutrons.

To see what protons are made of, we must view them on the subatomic scale. Particle accelerators have helped scientists do this. The Large Hadron Collider accelerates protons to near light speed before smashing them together. Proton dust can be observed this way. Physicists find that they are made of two forms of subatomic particles. These two forms included bosons and fermions. Protons have four vector bosons, one scalar boson, and twelve fermions.

Jacob aligns to the scalar boson called Higgs.

Jacob's four wives align to the four vector bosons called gluons, photons, Z bosons, and W bosons.

Jacob's 12 sons align to the twelve fermions called leptons and quarks. Each one of these fermions is said to also have a corresponding antiparticle. Those would be their wives.

The LORD's word is true; Jacob's descendants shall be as the dust of the earth.

The LORD absolutely gave this Bible to us. Trust GOD. (Psalm 97:9, 115:3, 148:13) (2 Chronicles 6:18) (1 Kings 8:10-13)

(Isaiah 2:5) "Come, descendants of Jacob, let us walk in the light of the LORD."

As protons accelerate through the Large Hadron Collider, they radiate photons, the quanta of light.

(Isaiah 9:2) "The people walking in darkness have seen a great light; on those living in the land of deep darkness a light has dawned."

This reveals that Jacob's family, the founders of the house of Shem, is the proton. To form a spiritual Earth, there must be spirit matter. The Kingdom also has an electron and a neutron.

Noah had three children, which began the families. Years later, a man named Abraham had children with three women. Each woman descended from a different son of Noah. Abraham then gave all of his children spiritual gifts to begin religions with. Here are the three religious families and their atomic alignments.

The family of Jacob is the Christians and the Jews. They are the atomic proton.

So far, electrons have not been able to be split into different parts. It is said that the Large Hadron Collider isn't powerful enough to do so. The family of Ham is the Baha'is and Muslims. Baha'i is a religion of unity. They believe that all religions are one. The Qur'an also speaks about remaining as one, like an electron. Here is a verse that expresses this.

(The Family of Imran Surah 3:103) "Hold fast to God's rope all together; do not split into factions. Remember God's favour to you: you were enemies and then He brought your hearts together and you became brothers by His grace; you were about to fall into a pit of fire and He saved you from it – in this way God makes His revelations clear to you so that you may be rightly guided." (M.A.S. Abdel Haleem Edition)

A neutron is made of three quarks. The family of Japheth is made of religions from three areas of philosophy. These areas include Indian philosophy, Oriental philosophy, and Middle Eastern philosophy. The Indian religious views are Hinduism, Buddhism, Jainism, and Sikhism. The Oriental religious views are Confucianism, Shinto, Falun Gong, Tenrikyo, Cheondoism, Cao Dai, and Taoism. The Middle Eastern religious views are Rastafarianism and Zoroastrianism. Rastafarianism is a form of Christian (Middle Eastern) view. The Middle Eastern philosophies are also simply categorized as other. The doctrines of the gods are considered neutrons, though they are not part of the central house of Japheth.

It is apparent why the families of Noah and Abraham have three parts. Without a complete atom, spirit matter would not have been achieved. In this setting, they form hydrogen. Stars are made of hydrogen. Unlike natural matter, Spirit matter doesn't decay. (Daniel 12:3) Let there be eternal light!

(Genesis 15:5) "Then He brought him outside and said, "Look now toward heaven, and count the stars if you are able to number them." And He said to him, "So shall your descendants be."

(John 8:12) "When Jesus spoke again to the people, he said, "I am the light of the world. Whoever follows me will never walk in darkness, but will have the light of life.""

(John 1:4-5) "In him was life, and that life was the light of all mankind. The light shines in the darkness, and the darkness has not overcome it."

Those are the atomic alignments of the three families. Each of the three also has their own psychological atomics, though. This is a little different. Their separate psychology is done with the Urim and Thummim. On the Urim, Shem is the proton mind and the electron heart. On the Thummim, Ham is the proton heart and the electron mind. Japheth is considered either the element of social hydrogen as in a protium isotope or the social neutron, on both. As a social neutron, Japheth completes a deuterium isotope.

When these two psyche atoms of the Urim and Thummim are brought together, the isotopes fuse to make molecular hydrogen gas H2. Each of the three families is then handed a personal psychological Urim. The Urim has eight outer sections. The center ninth section is their house. The house of Shem has the eight psychological proton sections. The house of Ham has the eight psychological electron sections. The house of Japheth has the eight psychological neutron sections. Together these five Urim functions form the living water molecule H_2O. (John 4:6-14, 7:38) (Revelation 7:17, 22:1, 22:17)

Jerusalem is Earth

Now that matter was formed, there was something for the earth to be constructed with. Next, we will observe Earth from a biblical perspective.

It has been thought that Jerusalem is a city in the Middle East. According to the Bible, Jerusalem is Earth. (Isaiah 66:13) (Galatians 4:25-26) We also find this fact in the book of Second Esdras. In these next two quotes. Jerusalem is referred to as a mother.

(GNT 2 Esdras 2:2-6) "Jerusalem, the mother who brought them into the world, says to them, "Go your own way, my children; I am now a widow left completely alone. I took great delight in bringing you up, but you sinned against the Lord God and did what I knew was wrong, so I mourned in deep grief when I lost you. What can I do for you, now that I am a widow and left completely alone? Go, my children, and ask the Lord for mercy.' "Father Ezra, I call on you to testify against these people as their mother has done, because they have refused to keep my covenant. Now bring confusion on them and ruin on their mother, so that they will have no descendants."

(GNT 2 Esdras 2:15-19) ""Mother Jerusalem, take your children in your arms. Guide their steps in safe paths; raise them with the same delight that a dove has in raising her young. I, the Lord, have chosen you. I will raise your dead from their graves because I recognize them as my people. Jerusalem, mother of these people, do not be afraid; I, the Lord, have chosen you. "I will send my servants Isaiah and Jeremiah to help you. At their request I have consecrated and prepared for you twelve trees, heavy with different kinds of fruit, twelve fountains flowing with milk and honey, and seven high mountains covered with roses and lilies. I will make your children very happy there."
In this third quote, Jerusalem is first revealed as the mother and then unveiled as the Earth.

(GNT 2 Esdras 10:7-9) "Jerusalem, the mother of us all, is overcome with grief and shame. You ought to be mourning for her and sharing the grief and sorrow of all of us. But you are mourning for that one son of yours. Ask the earth; let her tell you that she is the one who ought to be mourning for the vast multitudes of people that she has brought to birth. The woman you saw is Jerusalem, which you now see as a completed city."
To scientifically back this as fact, research has been included.

There is a man who figured out how to measure the circumference of Earth with the Bible. Basically, he began with the land size of the map in the Book of Ezekiel, divided by the size of the city in the same book, multiplied by the size of the city in the book of Revelation.

There were three measurements found for the Roman furlong: 606.25 feet, 606.5 feet, and 606.84 feet. He used a furlong of 606.84 feet to calculate this. The Apostle Paul would have used the same furlong.

Here are the calculations.

606.84 ft. x 3,000 furlongs / (5,280 feet / mi.) = 344.80 mi. (554.89)

325,000 / 4,500 * 344.80 mi. = 24,902 mi. (40,075 km)

His result was .170% more than the polar circumference and .0001% more than the equatorial circumference.

This information on the circumference of the earth didn't come from the author of this book. It was important and therefore was incorporated into the work.

Here is a link to the site that revealed this data.

https://www.pickle-publishing.com/papers/ezekiels-city-circumference-of-the-earth.htm

This measurement of Earth is scientific proof. The Bible is from the LORD. We know that Abraham inherits the entire Earth. (Isaiah 65:17, 66:22) (2 Peter 3:13) This planet is now to be formed with Spirit matter.

Timeline of Evolution

According to common interpretations of the Bible, the world was created somewhere near 4,000 B.C. This notion seemed to be proven wrong by modern science. Diverse species of humans have lived here. Millions of years ago there were dinosaurs. Long before that, microbial life was already thriving. Even before microbes, the earth was under construction. Who is correct? If the Bible is from GOD, and GOD is true, then why don't these timelines match?

Scientific data has helped estimate that the earth formed about 4,540,000,000 years ago. The earliest possible appearance of life is said to have been as far back as 4,280,000,000 years ago. Life on Earth has three parts. The three domains of life are known as prokaryotes (bacteria), archaea, and eukaryotes (eukarya). This aligns with the three families of Noah and Abraham.

Prokaryotes (bacteria). Shem family.

Archaea, like bacteria, are single celled. Ham family.

Eukaryotes (eukarya). Japheth family.

We must be able to prove that the Bible explains the evolutionary process of life. This is fairly easy to do, as long as science is accurate. We begin with comprehension of this verse. (2 Peter 3:8) According to the LORD, each day equates to a thousand years. The Qur'an and the Latter-Day Saint Book of Abraham both concur.

The length of a day slowly shifts over time. It has been said that a billion years ago, a day lasted about nineteen hours. Since days change, we must find the correct timeframe to reckon with.

A set of seals was produced to find the first day of the new era. They are the seals of Exodus and Revelation in Book 5 of this series. Those are then aligned to the aurora checkpoint seals in chapter 27 of Book 7 of this series. These two sets of seals reveal that the previous timeline ended on 12-21-2011. Day 1 A.G. was on 12-22-2011. We are now in the proper biblical timeline A.G.

To get the correct calculations, we use the number of days in a year between 12-22-2011 and 12-21-2012. Each year is said to be 365.2422 days. We multiply that by 1,000 to get the length of a year to the LORD. This means that each year in the biblical lineage timeline represents 365,242.2 years.

Two sites were used to match the biblical lineage dates. This first link is to the site used to get the timelines of science.

https://en.wikipedia.org/wiki/Timeline_of_the_evolutionary_history_of_life

This second link is to the site used to get the biblical timelines.

https://biblehub.com/timeline/

To align the biblical timelines correctly, we add about 2,011 years to the B.C. dates. Here is what we find.

The oldest fossil evidence of eukaryotes is from about 2,000,000,000 years ago. The Bible reveals that Adam was created about 6,011 years before day one A.G. We first multiply 6,011 by 365,242.2. The answer is that Adam was created about 2,195,470,864 years ago. Around then, eukaryotic cells were being formed. The creation of mankind began. The garden of Eden was planted. This could be about when eukaryotes first fell away from archaea. During that time the Vredefort impact structure was formed by a crater that hit earth.

At about 5,011 years before day one A.G., it was said that Cain killed Abel, and the generations passed from Adam to Noah. Using the same mathematics, this calculates to about 1,830,228,664 years ago. According to science, a new form of eukaryotic cells derived about that time. They were most likely the result of prokaryotes engulfing each other. (Judges 14:14) This is how we can see the forbidden fruit. They were basically eating each other. This is similar to cannibalism and eating meat on a microscopic level. Amoebas were eating one another, including their children. That would be the fruit of the loins of their own family trees. (Lamentations 2:20) (Deuteronomy 28:54-57)

During about 4,511 years before day one A.G., wickedness was said to have provoked GOD's wrath. There was a great flood, and life was separated into three groups. According to biblical math, this was about 1,647,607,564 years ago. Science concludes that around 1,600,000,000 years ago, protosterol biota eukaryotes formed. They lived in Earth's waterways. They have been considered the world's first predators. That would provoke GOD's wrath. This was also around the beginning of what is called the Mid Proterozoic Middle Ages of Earth. During that time, evolution almost stopped.

From about 4,101 and 3,817 years before day one A.G., Abraham, Isaac, and Jacob with his children were here. That was sometime between 1,497,858,262 and 1,394,129,477 years ago. Exceptionally well-preserved microfossils called Volyn biota have been dated to this time. During those days, eukaryotes are said to have begun evolving towards a stable level of diversity. This aligns with Abraham, Isaac, and Jacob, whom GOD employed to bring global peace through stable diversity.

From about 3,536 to 3,386 years before day one A.G., Moses and Joshua were here. That was sometime between 1,291,496,419 and 1,236,710,089 years ago. According to science, the earliest land fungi appeared.

Meiosis and sexual reproduction in eukaryotes were becoming a part of the evolutionary path. The type of intercourse that this aligns to is human intercourse with heavenly design. The Tabernacle and Law of Moses both had designs for male and female Priesthoods. Though they had been given the design, it was temporarily put away.

From about 3,035 to 2,942 years before day one A.G., David and Solomon were here. This was sometime between 1,108,510,077 and 1,074,542,552 years ago. It has been estimated that near those days, the first eukaryotes moved to land. These non-marine life forms became a new step in the evolution of plants. This aligns with when King David chose a plant-based diet. This truth was revealed by the fact that he was said to have taken the lamb. The lamb means veganism. The people were upset with David because they couldn't comprehend the light of his countenance.

Jesus came about 2,011 years before day one A.G. Using biblical math, this was about 734,502,064 years before 12-22-2011. The beginning of animal evolution was estimated to have begun about 750,000,000 years ago. Jesus brings evolutionary changes. Without these changes, life on Earth would cease to exist in a dead end. This timeline is correct.

Dinosaurs were here between about 230,000,000 and 66,000,000 years ago. This can be divided by 365242.2 to get 630 to 181 years ago. The Apocrypha mentions dinosaurs. Many dinosaurs were very large and tall compared to modern species. (2 Esdras 5:51-55) There are two types of children, both spiritual and natural. The religious dinosaurs lived between the years of about 1381 and 1830. That was the time of great religious powers and wars.

Homo (H. habilis) were the first humans that lived here about 2,500,000 years ago. Using the same biblical mathematics, we get about 6.844773139576971 years or 2,500 days. That is then subtracted from 12-22-2011 to get a date near February 15th 2005. This means that the current and new evolution was initiated near the beginning of 2005. An explanation on the initiation can be found in Book 1 of this series.

The LORD causes evolutionary changes. Angels administer on planets and help bring life. The Homo sapiens who lived here prior to 12-22-2011 became an outdated species. They were too concerned with monetary gain. They weren't considering future generations. Because of their design, they were destroying the Earth. It is time for evolution. We can revive our lives from the four winds (voices). The four winds are also known as the four riverheads of doctrine. (Ezekiel 37:1-10)

To complete this seal, we begin with 12-22-2011 as day one A.G. Calculating one year using the LORD's mathematics, we get about 365,243.2 B.C.E. That was when the Neanderthals met a great evolutionary period in their lives. That was around the emergence of sapiens. That is like a day in human evolution. (Ezekiel 4:6)

Age of the Earth

The estimated timeline of the Earth's formation is about 4,540,000,000 years ago. The book of Second Esdras states that the history of the world is divided into twelve periods. It says that the tenth part had arrived and already was halfway over. Only two and a half parts remain. (2 Esdras 14:11-12) That is the book of John the Baptist. If this is correct, the end of the world can be calculated from 1 A.D. We divide 4,540,000,000 by 9.5 to get 477,894,736.8421053. We then multiply that by 2.5 and subtract 2,011 years to get an estimated end date of 1,194,734,831 A.G. Scientists say that the sun will get hot enough to boil the oceans in about a billion years. Second Esdras is scientifically correct. Scripture states that the LORD will still be here taking care of us when that happens. (Revelation 21:1) (Isaiah 30:26)

According to Second Esdras, the sun eventually shines at night while the moon does during the daytime. This refers to the length of days changing over time. The orbit of the sun and moon will shift. (2 Esdras 5:4)

It seems that the Book of Jubilees may be important. It appears to be linked to the timelines of atmospheric and weather changes. The jubilees or lesser time periods, should be explored. If this is true, research will reveal it. Jubilee calculations may or may not align to exact dates. They may be in sequence of story instead of year.

(Book of Jubilees 50:13) "The man who does any of these things on the Sabbath shall die, so that the children of Israel shall observe the Sabbaths according to the commandments regarding the Sabbaths of the land, as it is written in the tablets, which He gave into my hands that I should write out for you the laws of the seasons, and the seasons according to the division of their days."

Based on the version by R.H. Charles (*Apocrypha and Pseudepigrapha of the Old Testament*, Oxford: Clarendon Press,1913) with the archaic word-forms modernized by Patrick Rogers

This science has been revealed so that others can complete it. (Matthew 10:26) (Mark 4:22) There are more timelines to use with the biblical lineage. A year can be converted to a day. (Ezekiel 4:5) That conversion can be used for psychological growth in a child. Likewise, a day can mirror a year. There are also other conversions, such as weeks to years. (Daniel 9:2, 9:24-27) With correct conversions, the evolution of humanity can be linked properly. Whatever is revealed may be in an abstract. Using metaphors and the meanings of names, the entire story will unfold.

12-22-2011 is now precisely day one A.G. The Bible can reflect and expand time in both forward and reverse. The future can be seen by viewing the past. It was designed in fractal essence. Since the beginning of time is now before it was yesterday, the LORD can absolutely have created HIMSELF.

Timeline of Psychological Generations

Each name in the Bible has a definition or meaning that coincides with the behavior of that person. This is not a coincidence. Think of a bloodline in the Bible. Each person is a process of thought or behavior. Each child's name, by definition, is the outcome of the previous thoughts (parents). The thought process of Solomon was the outcome of the unified thought processes of David and Bathsheba. Each personality generation may go several ways. David had twenty children with seven women.

Here is an abstract to explain this. Let's say that there was a dad with blue skin. His name was Blue. He had a wife named Bicycle. Together they had a child named Riding. Now Riding had three children named Fast, Fun, and Crash. Crash always teamed up with Fast, attempting to make war with Fun. Because of this, Blue blessed Fun, while Fast and Crash got hurt.

Following some lineages of thought proves to render psychological dead ends. Jesus' lineage (thought process) proves eternal life. The generational psychology of Jesus Christ provides an outcome with the resurrection of righteous thoughts and intentions.

Timeline of Astronomical Events

The generations and stories in the Bible are a map of astronomical events in the stars. When timelines are calculated correctly, the universe aligns. Because of this, the Bible can properly connect astrology.

Seven Days of Religious Creation

The LORD GOD had planted a garden eastward in Eden. This means that the Garden of Eden was a social cognitive/emotional unity. (Genesis 2:8)

Out of the ground HE made every tree grow that is pleasant for food. The trees of life and knowledge of good and evil were there. (Genesis 2:9)

There was a river sourced in Eden. The river is a spiritual doctrine graphed into Source Frequency. The doctrinal connection is mentioned in chapter 1 of this book. That single river parted and became four riverheads. The four rivers are the four original regions of doctrine. (Genesis 2:10-14) These four doctrinal regions are well defined in Book 8 of this series.

Pishon – Shem – Jews and Christians.
Gihon – Ham – Muslims and Baha'is.
Hiddekel – Japheth – Alternate religions other than the Orientally founded.
Euphrates – Oriental Japheth – Oriental alternate religions.

Eventually, the two riverheads of Japheth merged when Buddhism entered the Orient lands. The merging is mentioned in the book of Revelation. (Revelation 16:12) Spiritually speaking, the Euphrates dried up because it became one with the Tigris. It is safer that they joined. The other two religion riverheads have been known to fight against them with a passion.

This next gnostic verse reveals that these many religions are set into the root (Source). The mathematics in chapter one of this book reveals the root.

(Gnostic The Interpretation of Knowledge) "Therefore they are lovers of abundant life. And each of the rest endures by his own root. He puts forth fruit that is like him, since the roots have a connection with one another and their fruits are undivided, the best of each. They possess them, existing for them and for one another. So let us become like the roots, since we are equal"

CODEX XI Translated by John D. Turner Selection made from James M. Robinson, ed., The Nag Hammadi Library, revised edition. HarperCollins, San Francisco, 1990.

The LORD GOD placed Adam in HIS garden to take care of it. Adam was allowed to eat from every tree except for the tree of the knowledge of good and evil. (Genesis 2:15-17) This was a spiritual garden. The edible foods signify different religious doctrines. The tree of knowledge and its forbidden fruit was meat and the world's governing authorities.

All religions were connected in masculine social cognition and feminine social emotion. Once Adam had eaten those things, the two social unities were broken apart. Religions began fighting and disagreeing. They went blind to the truth. This happens because the worldly governing forces need problems. They discreetly cause universal contentions. They then solve problems in exchange for honors resembling worship. By using force to resolve issues, they inconspicuously plant seeds of destruction. They are trained as professional predators. Their path is the enemy of all spirituality.

To properly see how these religions were given to us, we begin with the seven days of creation. This is another veil of the Ark of the Testimony. Each day of religious formation represents 1,000 years. (2 Peter 3:8) The spiritual doctrines were manifested during a 7,000-year era. The 7,000 years have a pre-era and a post-era.

The pre-era can be seen in the seal of Ephesus. We notice that something has already been accomplished when the first of seven seals begins. Whatever was accomplished was done during the pre-era. (Revelation 2:1-3) After the seventh day of creation, an effectual door opens to begin a new first day for a new era. The new era begins on the eighth day of the previous creation.

Here is how the biblical eras are aligned. Three eras of seven thousand years are mentioned in this list. The first and third eras only have one day mentioned. The second era is complete. Though there are many more, these eras are cited as one, two, and three.

Day 7 was pre-religion to 4989 B.C.
Day 1 was from 4990 to 3989 B.C. (This is also the 8th day of the first era.)
Day 2 was from 3990 to 2989 B.C. Adam most likely received written language about 3502 B.C.
Day 3 was from 2990 to 1989 B.C.
Day 4 was from 1990 to 989 B.C.
Day 5 was from 990 B.C. to 10 A.D.
Day 6 was from 11 to 1010 A.D.
Day 7 was from 1011 to 2010 A.D.
Day 1 was from 12-22-2011 to 3010 A.D. (This is also the 8th day of the second era.)

There are three Adams mentioned in the Gnostics. The first Adam was created at the end of the seventh day of the first era. He then appeared to the people during the first day of the second era. He brought spirits to humanity. The second Adam, Jesus, was created at the end of the fifth day of the second era. He then appeared to the people on the sixth day of the second era. He brought souls and provided an afterlife to the people. The second Adam had a spirit because of the works of the first Adam.

The Epistle of Barnabas reveals that everything came to an end with Jesus. The LORD had shut the veil of the Bible so that people couldn't fully comprehend it. Again, the veil of the Bible has opened during the first day of the third era. Here is a quote from the Epistle of Barnabas.

(Barnabas 15:4) "Give heed, children, what this meaneth; He ended in six days. He meaneth this, that in six thousand years the Lord shall bring all things to an end; for the day with Him signifyeth a thousand years; and this He himself beareth me witness, saying; Behold, the day of the Lord shall be as a thousand years. Therefore, children, in six days, that is in six thousand years, everything shall come to an end.

Translated by J.B. Lightfoot.

According to alignment and prophecy, the third Adam would be created at the end of the seventh day of the second era. He would then appear to the people during the first day of a third era. The first day of the third era is also called the eighth day of the second era. The third Adam is said to bring scientific information and a living law for the people. Without that law, humans would go extinct. He is sent to save Earth before the people destroy it. The third Adam also has a spirit and soul because of the works of the first and second Adams. He comes to connect heaven and Earth. The third Adam brings the Tree of Life and refuses the forbidden fruit. Here is the Gnostic alignment.

(Gnostic On The Origin of the World) "Now the first Adam, (Adam) of Light, is spirit-endowed (pneumatikos), and appeared on the first day. The second Adam is soul-endowed (psychikos), and appeared on the sixth day, which is called Aphrodite. The third Adam is a creature of the Earth (choikos), that is, the man of the law, and he appeared on the eighth day [...the] tranquility of poverty, which is called Sunday. And the progeny of the earthly Adam became numerous and was completed, and produced within itself every kind of scientific information of the soul-endowed Adam. But all were in ignorance."

"The Untitled Text" CODEX XIII Translated by Hans-Gebhard Bethge and Bentley Layton Selection made from James M. Robinson, ed., The Nag Hammadi Library, revised edition. HarperCollins, San Francisco, 1990.

Before the First Day – Weight
– This was before 4989 B.C. during day seven of the previous era. (Job 10:22) (Psalm 139:11-12) (Barnabas 15:4) (Genesis 1:1-2) (2 Esdras 6:35-37)

GOD had not yet given us our modern path. Religion was without form and void, and darkness was on the face of the deep. Some people had spirituality within types of utter darkness. They would have used animal symbolism. That included acting like them. There were possible religious activities at sites such as Göbekli Tepe. Spirituality included violence and sacrifices. Adam was born.

Evening and Morning of the First Day – Negative – This was between 4990 and 3989 B.C. (Genesis 1:3-5) (2 Esdras 6:38-40) (Moses 2:1-5) (Abraham 4:1-5)

GOD created a spiritual light. This spirituality was found in kindness and love. It was separated from the ways of murder, sacrifice, and violence. (Job 12:22) (Psalm 112:4) (Isaiah 42:16) (Luke 1:79, 11:35-36) (John 1:5) (Romans 2:19) Adam and Eve were chosen from the people to lead a new spirituality. GOD warned Adam that he wasn't to rejoin the ways of the people in darkness. Their ways included eating meat and fighting for honor. During those days, there wasn't yet a complete writing ability.

Evening and Morning of the Second Day – Positive – This was between 3990 and 2989 B.C. (Genesis 1:6-7) (2 Esdras 6:41) (Moses 2:6-8) (Abraham 4:5-8)

GOD provided a doctrine from heaven called the Urim. The waters (spirits) above administered to the waters (spirits) below. Those who believed in and followed GOD joined the waters above. Those who disbelieved in and denied GOD remained with the waters below.

Sometime around 3502 B.C., a written language was given to the people. Adam may have been the first with the ability to write. The Urim and Thummim was a doctrine from heaven. Evidence of the Urim can be seen in the Kesh Temple Hymn. More about this can be found in Book 7 of this series.

The story of Cain slaying Abel aligns to the way that the people of darkness acted towards those who chose the light. Simply behaving in a godly way was an offense to the darkness. This is seen in animalistic behavior. Some animals will kill others in a pack for showing weakness. In the animal kingdom, dominance rules. Men who have been trained to socialize with semi-wild animals such as lions must be dominating with confidence. The light of spirituality leads with thoughtfulness and love. It can be mistaken as a weakness. Those leading with dominance didn't like the light. They would kill those who would reveal domination and force as incorrect leadership.

(Second Book of Adam and Eve 19:1-4) "Then God revealed to him again the promise He had made to Adam; He explained to him the 5500 years, and revealed unto him the mystery of His coming upon the earth. And God said to Jared, "As to that fire which thou hast taken from the altar to light the lamp withal, let it abide with you to give light to the bodies; and let it not come out of the cave, until the body of Adam comes out of it. But, O Jared, take care of the fire, that it burn bright in the lamp; neither go thou again out of the cave, until thou receivest an order through a vision, and not in an apparition, when seen by thee. "Then command again thy people not to hold intercourse with the children of Cain, and not to learn their ways; for I am God who loves not hatred and works of iniquity."

The Forgotten Books of Eden, by Rutherford H. Platt, Jr., [1926], at sacred-texts.com

The ways of Cain are the ways of the world's governments. Religious leaders are not to use or promote guns, bombs, war, prisons, and many other ways of theirs.

Evening and Morning of the Third Day – Multiplicative – This was between 2990 and 1989 B.C. (Genesis 1:8-10) (2 Esdras 6:42-44) (Moses 2:7-12) (Abraham 5:9-13)

GOD gathered the waters (spirits) under the heavens together. Dry land called earth appeared. Those who had listened to GOD became the earth. (Psalm 37:11) (Matthew 5:5) The people called Earth began yielding many types of spirituality. That is what it means by grass, herb that yields seed, and fruit trees yielding fruit with seeds. (Proverbs 11:30, 12:12) (Isaiah 3:10, 32:16) (Ephesians 5:9) (Philippians 1:1) (James 3:18)

Those who remained predators were called the sea. They had doctrines that were not from GOD. (Isaiah 57:27) It was estimated that one out of every seven people fought against the LORD. This is mentioned in Second Esdras. (2 Esdras 6:42-44) According to statistics, this scientifically aligns to the fact that even today, one in 7, or about 15% of people, have worked for the government. (2 Corinthians 6:17) Information on the percentages is found in this next link.

https://www.dailysignal.com/2024/10/10/government-employees-exceed-population-of-florida/

During these years there was a great flood. There would be many false religious beliefs, so GOD intervened. A covenant was made that the Urim and Thummim would be grafted into doctrines. Doctrines designed using a Urim were kept alive. The eight souls that were saved in Noah's Ark aligned with the eight outer sections of the Urim. The eight souls were Noah and his children with their wives. (1 Peter 3:18-22) Anything that didn't align to the Urim would drown in the flood. The Urim is the rainbow system. To extend the covenant, Abraham was aligned to the Urim with his eight children and three wives. (Luke 1:55)

Evening and Morning of the Fourth Day – Gravity & Pressure – (Genesis 1:14-19) (2 Esdras 6:45-46) (Moses 2:13-19) (Abraham 5:14-19) This was between 1990 and 989 B.C.

The covenant of Noah and Abraham was passed to Jacob. The Ark of the Testimony was placed into the covenant of the Urim. We can see this because Jacob had four wives and twelve children. There are twelve parts to the Ark of the Testimony. You can reference Chapter 1 of this book. Four wives allowed for a new Urim alignment with the four prime colors. It also sealed a doctrinal foundation for the four religious riverheads of Eden. (Psalm 105:6) (2 Corinthians 11:22) (Hebrews 2:16)

The sun, moon, and stars mean dad, mom, and children. These are symbols of the parents and children of the Ark of GOD. (Genesis 37:9-10) Moses and the Hebrews were blessed with the Ark of GOD through Jacob.

The covenant of the Urim was then passed on to King David. We can see this because he had nine women with him. Their names were Ahinoam, Abigail, Bathsheba, Maacah, Haggith, Abital, Eglah, Michal, and Abishag. To concur with the Urim directions, Michal was drawn to represent the counsel position of density. David gained the throne by having the full Urim. The throne will be spoken of in Chapter 5 of this book.

The greater and lesser lights were created along with the stars. This means that GOD gave us Hinduism. The new doctrine was a covenant-linked foundation for astrology. Hinduism was designed with the Urim. Book 7 of this series reveals proof of this truth.

GOD gave us the Oracle during these thousand years. (Acts 37-38) The Oracle is a clean language used to disguise doctrinal meanings. (Zephaniah 3:8-9) (Romans 3:1-4) (Hebrews 5:12) More about this can be found in Chapter 5 of this book.

There were many spiritual doctrines besides Hinduism throughout the world. Most of them spoke of gods. Those doctrines could be converted using the Oracle.

Evening and Morning of the Fifth Day – Division – (Genesis 1:20-23) (2 Esdras 6:47-52) (Moses 2:20-22) (Abraham 5:20-23) This was between 990 B.C. and 10 A.D.

GOD made the sea creatures and the birds of the air. Those against GOD were called the seas. The sea creatures align to miscellaneous doctrines of the gods. These doctrines were designed and placed into those who were against GOD. The governing forces were called Leviathan. This great sea creature would play with the doctrines of the gods. Leviathan cannot be beaten through carnal fighting. The government thrives on war. They manifest battles for honor. We are warned not to stand up and physically fight the government.

They love war. They kill for sport. (Job 41:1-34) Only the LORD can punish them. (Psalm 74:14, 104:26) (Isaiah 27:1) All religions are warned not to partake in any of their ways.

Most of the doctrines of the gods were elemental for astrological graphing. Putting these together is only allowed with the Urim. They are to be translated with the Oracle. The reason that they were written with the Oracle was because the Leviathan destroys love. Governing forces are threatened by righteousness. With the Oracle, writers could hide the true meaning. Doctrines were disguised to look like they were aligned to animalistic nature. The Bible was written in Oracle so that Solomon and Jesus could get it delivered to the world. They would have been stopped otherwise.

The birds align to scriptures specifically made for heavenly righteousness. They include Buddhism, Confucianism, Taoism, Jainism, Zoroastrianism, and Shinto. These religions were also called the Behemoth. (2 Esdras 6:48-52) (Job 40:15-24)

Shinto was designed as separate from the sea creatures, although Leviathan would eventually be tempted to play there. In fact, governing forces would ultimately attempt to indoctrinate all religions into their warring ways. When they do this, the religions become military-minded. That is why war broke out in heaven. (Revelation 12:7) We are to fight the good fight. That means to war in righteousness. We cannot have a cold, hard heart like Leviathan. Listen to everything that people say. We are not to physically fight to solve problems. We cannot use guns to solve gun problems. (Revelation 19:11)

Evening and Morning of the Sixth Day – Volume & Density – (Genesis 1:24-31) (2 Esdras 6:53-54) (Moses 2:23-31) (Abraham 20:24-31) This was between 11 and 1010 A.D.

GOD created the living creatures according to their kinds. Within this veil, those creatures represent religions. Genesis mentions the cattle, creeping things, and beast in one order, right before reversing that order. That's because religious families weren't planted in the same order as their physical births. (Genesis 1:24-25)

Japheth had previously been initiated during days four and five. Now Ham and Shame were properly founded. The Orthodox and Catholic Churches were for Shem. The Muslims were for Ham. The three families of Noah and Abraham were brought forth.

During the sixth day, man was made in the image of GOD. This is about the second Adam, known as Jesus. Though He was born on the fifth day, He appeared on the sixth. Jesus was given dominion over all religions. (Genesis 1:26-31) The Hebrew Oracle had been completed. The Bible and many other Hebrew writings had a code for awakening all religions. The Word of GOD was together, though it wasn't time for them to wake. (John 1:1-18) With this Oracle, all religions could be graphed together properly.

The Seventh Day – Weight

– This was between 1011 and 2010 A.D. (Genesis 2:1-25, 3:1-6) (2 Esdras 6:55-59) (The Believers Surah 22:12-14) (Moses 3:1-25) (Abraham 5:1-21)

The seventh day of creation was a Sabbath rest. This means that the Oracle was put to sleep. The people didn't comprehend what the Bible meant. They couldn't translate the meanings. Certain people were spiritual enough to be allowed some information. Most of it was put away. Without proper comprehension of scripture, each man's word became his own oracle. (Jeremiah 23:33-38)

During the seventh day, the religions were sanctified together as one. Many more religious houses were brought to the people. These new portions of Shem, Ham, and Japheth were Sabbath offerings. Through the Spirit, Jesus brought these offerings to heal the people. Sikhism, Lutheran, Baptist, Presbyterian, Methodist, Tenrikyo, Baha'i, Southern Baptist, Cheondoism, Seventh Day Adventist, Pentecostal, Cao Dai, Rastafarianism, and Falun Gong were offered. (Luke 13:10-17)

The sanctification of religions was happening during the days near February 15th, 2005. During that time, GOD used successive events to call on the author of this book for delivering information. The events are mentioned in Book 1 of this series.

When reading chapter one of Genesis, you may notice that each day is from evening until morning. Evening till mourning is the night. This means that all religions were in darkness. They had been under the sway of non-spiritual governing forces. The seventh day is when the light of religion is sanctified. The people are then called to a greater light. (1 Thessalonians 5:5)

The sanctification was placed into the author of this series so that it could be revealed. Book 7 of this series has a galactic map. That map was made by stripping the readings from several astrological calendars. The readings were then redesigned using the Ark of the Testimony with the Urim and Thummim. Doing this changed the sun, moon, and stars. This information was mentioned in the Epistle of Barnabas.

(Barnabas 15:5) *"And He rested on the seventh day. this He meaneth; when His Son shall come, and shall abolish the time of the Lawless One, and shall judge the ungodly, and shall change the sun and the moon and the stars, then shall he truly rest on the seventh day."*

Translated by J.B. Lightfoot.

The governing forces are the lawless one. What they call laws are different from land to land. They are not moral. And since they were confounded as not acceptable by GOD, the carnal governments are lawless. (2 Thessalonians 2:8-9) The new Law directly from heaven is in book 11 of this series.

The Eighth Day – Negative – (Barnabas 15:8-9) This was between 12-22-2011 and 3010 A.D.

After the seventh day was completed, the eighth day began a new era. During the eighth day, GOD begins to harvest the doctrines. Heaven and Earth are being grafted together. The new era began on 12-22-2011. The correct Sabbath and calendar were aligned with scientific proof. Saturday is the Sabbath.

To find the correct day one A.G., a set of seals was produced. The seals of Exodus and Revelation are in Book 5 of this series. Those are then aligned to the Aurora Checkpoint Seals in chapter 27 of Book 7 of this series. These sets of seals reveal that the previous era ended on 12-21-2011. The next day was 1 A.G. We are now on proper biblical timeline.

The correct Sabbath, Saturday, was realigned on the eighth day, which is the beginning of the new era. The eighth (first) day is also a special day, for that was when Jesus was resurrected. Jesus had been dead for 3,000 years according to the days of Solomon. The Bible states that he was the son of the LORD. (1 Chronicles 28:6) After Solomon, the temple (body of GOD's people) was dismantled. (John 2:19-22) Jesus had been dead for 2,000 years according to the Gospels. Jesus was a person. Jesus is also the lineage of King David. (Revelation 22:16) Sunday sabbaths are not accepted by GOD. Jesus was here on the 6th day, and the Sabbath gets set on the eighth day (eighth thousand years). It was thought that the Sabbath was to be shifted to Sunday. The doctrine meant that Sabbath would be properly set when the eighth day begins a new era.

(Barnabas 15:8-9) "Finally He saith to them; *Your new moons and your Sabbaths I cannot away with.* Ye see what is His meaning; it is not your present Sabbaths that are acceptable [unto Me], but the Sabbath which I have made, in the which, when I have set all things at rest, I will make the beginning of the eighth day which is the beginning of another world. Wherefore also we keep the eighth day for rejoicing, in the which also Jesus rose from the dead, and having been manifested ascended into the heavens."

The Bible was long thought to have been false. Combining scientific proof, the Ark of the Testimony, and the Urim, the Word of GOD (Jesus) has now been resurrected. The Bible is the Word of GOD. The LORD gave many doctrinal writings to us. (1 Corinthians 9:19-23) (John 16:25)

(Gnostic The Teachings of Silvanus) "Furthermore, it is difficult to comprehend him, and it is difficult to find Christ. For he is the one who dwells in every place, and also he is in no place."

(Gnostic The Teachings of Silvanus) "Open the door for yourself that you may know the One who is. Knock on yourself that the Word may open for you. For he is the Ruler of Faith and the Sharp Sword, having become all for everyone because he wishes to have mercy on everyone."

CODEX VII Translated by Malcolm L. Peel and Jan Zandee Selection made from James M. Robinson, ed., The Nag Hammadi Library, revised edition. HarperCollins, San Francisco, 1990.

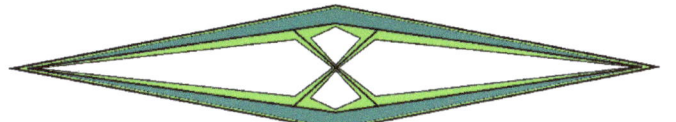

Chapter 5
The Oracle

Walking on Water

The Oracle is a clean language used to disguise doctrinal meanings. (Zephaniah 3:8-9) It was designed to help hide what doctrines meant. This aided in creating heavenly scriptures without being noticed by the governing forces of the world. World leaders were known to attack anything that didn't support their animalistic ways. It was safe to use because only through the Spirit could it be translated. Those who worked for the governments couldn't interpret the language.

The second ability of the oracle was that it could translate spiritual doctrines. By using the Oracle, the languages of many scriptures could be graphed together as one. (Ezekiel 43:2) (Revelation 1:15, 14:2)

To walk on water means to walk on doctrine. Notice that Jesus walks on the sea. (Matthew 14:25-26) The sea usually represents the forceful governing authorities of the earth. (Isaiah 57:20) Consider a new church. The attendants begin researching the Bible. Without the Oracle, they usually and almost instantly sank in the sea. They would read verses about how they are to obey the governing authorities. They learn that the authorities were established by GOD. Here are verses that teach this. (Romans 13:1-7) (1 Timothy 2:1-3) (Titus 3:1) (Hebrews 13:17) (1 Peter 5:5) When the Bible is translated correctly, it means to obey Noah, Abraham, Isaac, Jacob, Moses, King David, Jesus, and others that GOD had approved. Those are the authorities established by GOD.

Jesus abolished the law of commandments contained in the ordinances. The first law became obsolete. Here are verses that speak about this fact. (Ephesians 2:15) (Colossians 2:14) (Hebrews 7:18, 8:13, 10:1, 12:27) The people thought that GOD destroyed the Law of Moses. The New Testament warns that death reigned from Adam to Moses. The Law of Moses was a fallen law. It became a worthless Law which would lead to the fall of mankind. There was more to it, though. First, Moses' Law was designed to mirror the outrageous laws of the forceful governments of humanity. Their laws cause the fall of mankind. Theirs are also designed to attack GOD's people. (2 Corinthians 11:23) (Revelation 1:10) (Colossians 2:20-23) The real Law of Moses, behind the veil, is a living law of the Savior. It causes the reuniting of humanity with GOD.

To comprehend the veil of Moses, one must realize that it was written in an Oracle. (Exodus 34:33-35) (2 Corinthians 3:13-15) This next portion has many verses from the Epistle of Barnabas. Barnabas was another surname of Saint Peter, to whom Jesus gave the keys to. Barnabas hints at what the Law of Moses means. This is the record of when Saint Peter walked on water. (Matthew 14:29)

(Barnabas 2:6) "These things therefore He annulled, that the new law of our Lord Jesus Christ, being free from the yoke of constraint, might have its oblation not made by human hands."

(Barnabas 3:6) "To this end therefore, my brethren, He that is long-suffering, foreseeing that the people whom He had prepared in His well-beloved would believe in simplicity, manifested to us beforehand concerning all things, that we might not as novices shipwreck ourselves upon their law."

(Barnabas 9:8) "He who placed within us the innate gift of His covenant knoweth; no man hath ever learnt from me a more genuine word; but I know that ye are worthy."

(Barnabas 10:1) But forasmuch as Moses said; *Ye shall not eat seine nor eagle nor falcon nor crow nor any fish which hath no scale upon it*, he received in his understanding three ordinances."

(Barnabas 10:2) "Yea and further He saith unto them in Deuteronomy; *And I will lay as a covenant upon this people My ordinances.* So then it is not a commandment of God that they should not bite with their teeth, but Moses spake it in spirit."

(Barnabas 10:3) "Accordingly he mentioned the swine with this intent. Thou shalt not cleave, saith he, to such men who are like unto swine; that is, when they are in luxury they forget the Lord, but when they are in want they recognize the Lord, just as the swine when it eateth knoweth not his lord, but when it is hungry it crieth out, and when it has received food again it is silent."

(Barnabas 10:4) "*Neither shalt thou eat eagle nor falcon nor kite nor crow. Thou shalt not, He saith, cleave unto, or be likened to, such men who now not how to provide food for themselves by toil and sweat, but in their lawlessness seize what belongeth to others, and as if they were walking in guilelessness watch and search about for some one to rob in their rapacity, just as these birds alone do not provide food for themselves, but sit idle and seek how they may eat the meat that belongeth to others, being pestilent in their evil-doings.*"

(Barnabas 10:5) "*And thou shalt not eat,* saith He, *lamprey nor polypus nor cuttle fish* . Thou shalt not, He meaneth, become like unto such men, who are desperately wicked, and are already condemned to death, just as these fishes alone are accursed and swim in the depths, not swimming on the surface like the rest, but dwell on the ground beneath the deep sea."

(Barnabas 10:6) "*Moreover thou shalt not eat the hare.* Why so? Thou shalt not be found a corrupter of boys, nor shalt thou become like such persons; for the hare gaineth one passage in the body every year; for according to the number of years it lives it has just so many orifices."

(Barnabas 10:7) "Again, *neither shalt thou eat the hyena;* thou shalt not, saith He, become an adulterer or a fornicator, neither shalt thou resemble such persons. Why so? Because this animal changeth its nature year by year, and becometh at one time male and at another female."

(Barnabas 10:8) " Moreover He hath hated the weasel also and with good reason. Thou shalt not, saith He, become such as those men of whom we hear as working iniquity with their mouth for uncleanness, neither shalt thou cleave unto impure women who work iniquity with their mouth. For this animal conceiveth with its mouth."

(Barnabas 10:9) "Concerning meats then Moses received three decrees to this effect and uttered them in a spiritual sense; but they accepted them according to the lust of the flesh, as though they referred to eating.

(Hebrews 5:12-15, 6:5) (2 Corinthians 3:14-15)

(Barnabas 10:10) "And David also receiveth knowledge of the same three decrees, and saith; *Blessed is the man who hath not gone in the council of the ungodly*--even as the fishes go in darkness into the depths; *and hath not stood in the path of sinners*--just as they who pretend to fear the Lord sin like swine; *and hath not sat on the seat of the destroyers*--as the birds that are seated for prey. Ye have now the complete lesson concerning eating."

(Barnabas 10:11) "Again Moses saith; *Ye shall everything that divideth the hoof and cheweth the cud.* What meaneth he? He that receiveth the food knoweth Him that giveth him the food, and being refreshed appeareth to rejoice in him. Well said he, having regard to the commandment. What then meaneth he? Cleave unto those that fear the Lord, with those who meditate in their heart on the distinction of the word which they have received, with those who tell of the ordinances of the Lord and keep them, with those who know that meditation is a work of gladness and who chew the cud of the word of the Lord. But why that which divideth the hoof? Because the righteous man both walketh in this world, and at the same time looketh for the holy world to come. Ye see how wise a lawgiver Moses was."

THE EPISTLE OF BARNABAS – Translated by J.B. Lightfoot.

Those who wrote the scriptures taught that the carnal rulers of earth were destroying everything. They still are today. All scriptures are about getting rid of their ways. Government opinion was leading the religions. Officers used force and honor to take control by seduction. There were officers, who are enemies of GOD, in priesthood positions. The Bible clearly opposed this.

(NKJV Ephesians 6:12) "For we do not wrestle against flesh and blood, but against principalities, against powers, against the rulers of the darkness of this age, against spiritual *hosts* of wickedness in the heavenly *places.*"

Their desire isn't to help the people; they desire power. That is why, regardless of what they are doing, they are wrong. They then pass their thought processes onto the religions. Through their ways, the religions learn not to get along with one another. These Gnostic quotes concur.

(Gnostic Tripartite Tractate) "As they brought forth at first according to their own birth, the two orders assaulted one another, fighting for command because of their manner of being. As a result, they were submerged in forces and natures in accord with the condition of mutual assault, having lust for power and all other things of this sort. It is from these that the vain love of glory draws all of them to the desire of the lust for power, while none of them has the exalted thought nor acknowledges it."

CODEX I Translated by Harold W. Attridge and Dieter Mueller Selection made from James M. Robinson, ed., The Nag Hammadi Library, revised edition. HarperCollins, San Francisco, 1990.

(Gnostic The Book of Thomas the Contender) "The savior said, "Truly, as for those, do not esteem them as men, but regard them as beasts, for just as beasts devour one another, so also men of this sort devour one another. On the contrary, they are deprived of the kingdom since they love the sweetness of the fire and are servants of death and rush to the works of corruption. They fulfill the lust of their fathers.

CODEX II Translated by John D. Turner Selection made from James M. Robinson, ed., The Nag Hammadi Library, revised edition. HarperCollins, San Francisco, 1990.

We cannot physically fight them. They are a barrier. If we were to carnally attack them to win, then we are not ready to pass that barrier anyhow. The direction is to not participate in their ways.

When Jesus appeared as a real person, His direction was to unveil the Old Testament Law and renew it. He said not to join the bondage of the laws of the world. The governments got rid of the truth. When the scriptures were handed to the people, they simply had no Oracle. Jesus, who was also the Word of GOD, was crucified. He didn't die, though; the people did instead. (Romans 6:5-10) (2 Timothy 2:11-13) (1 Thessalonians 4:14)

(Gnostic The Second Treatise of the Great Seth) "For my death, which they think happened, (happened) to them in their error and blindness, since they nailed their man unto their death. For their Ennoias did not see me, for they were deaf and blind. But in doing these things, they condemn themselves. Yes, they saw me; they punished me. It was another, their father, who drank the gall and the vinegar; it was not I. They struck me with the reed; it was another, Simon, who bore the cross on his shoulder. I was another upon Whom they placed the crown of thorns. But I was rejoicing in the height over all the wealth of the archons and the offspring of their error, of their empty glory. And I was laughing at their ignorance. And I subjected all their powers. For as I came downward, no one saw me. For I was altering my shapes, changing from form to form. And therefore, when I was at their gates, I assumed their likeness."

CODEX VII Translated by Roger A. Bullard and Joseph A. Gibbons Selection made from James M. Robinson, ed., The Nag Hammadi Library, revised edition. HarperCollins, San Francisco, 1990.

Jesus veiled His doctrine and made it look like it matched the corrupt leaders. It gathered together those who in their hearts would already go that way. Anyone who picked up the Bible had a choice. They would either join the carnal mindset or see in their hearts that something wasn't right. This helped filter the people. Without the Oracle, the Christians sank in the sea. They were unable to walk on water.

Jesus said to get swords. The people came back with two. (Luke 22:36-38) One sword represents a real sword. Those who read the Bible usually see that the Hebrews killed the Canaanites. Many of GOD's chosen men killed enemies in battle. They see that and grab the real sword. Those who grab that sword sink in the sea. They are all invited to the LORD's great sacrifice. They are the sacrifice. (Joel 3:10-17) (Isaiah 34:5-12) (Zephaniah 1:7-9) The other sword represents the unveiled and sharpened Word of GOD. Those who grab this sword survive Noah's flood. They don't drown in the sea. (Isaiah 2:3-4)

Translating the Oracle

The Oracle is a clean language used to write the Bible. It has several veils. The first veil helps translate meanings and directions given by GOD. It has things such as Babylon meaning worldly governments, etc. The story of Jacob and Edom is from the outer veil. The inner veil is set into a paradisiacal view and can be considered universal. (Jeremiah 17:15-16) We pass through the outer veil before getting to the inner veil.

To better comprehend the Oracle, we go to the story of Esau, the son of Isaac. Esau was the firstborn. When it came time to bless Esau with the Abrahamic covenant, his brother Jacob took it. To do this, Jacob dressed like his brother. Their dad was nearly blind. Isaac gave the disguised Jacob the blessing instead. (Genesis 27:1-29) Here is what this means. The veiled Bible is Jacob's disguise. It was designed to look like it agreed with the governmental ways. It has stories such as the LORD told Israel to kill all of the Canaanites. Moses tells people to kill people with stones. It looks like it fits right in with the military mindset. All of that is outside of the veil. The Bible was dressed like the Edomites who live by the carnal sword. It was done to take the blessing from Esau. (Genesis 27:30-40)

There is more to the story of Edom. Esau means a previous species. Jacob means the new species. Here is doctrinal evidence of this fact. (2 Esdras 6:7-10) What this all means is that the Christians were Edom. Jacob is the Bible. The Word of GOD births a new species within the people of Edom. The Bible had looked like it fit right in with the Edomites. In this veil, Isaac and his wife were the parents of the Christians. They were blind and couldn't translate the Bible. The Christians lost their blessing and must transition to Jacob for their blessing to return to them.

The Edomites lost their blessing by adhering to a carnal military mindset. They had also traded their birthright for the flavor of meat. Here are the first rules for receiving their rights and blessings. All religions must disassociate from carnal government. There are not to be any carnal officers or governmental workers in any leadership positions throughout the religions. They are to enter vegetarianism as a dietary minimum. Vegetarianism is to be used as a step towards veganism. Those are the rules from heaven.

Law of Moses Keys

Untranslated, the Law of Moses would cause a fall of mankind. When translated, it becomes the Living Law of Jesus. It supports and saves the world. There are many keys used in translating the Law of Moses. Most of them are revealed in Book 11 of this series. The Oracle allows comprehension of the Law.

There are sacrifices and offerings written in the Law of Moses. They represent tithe-type contributions. They also often signify educational lessons. To sacrifice a bull means that a presiding Apostle or Disciple gives a message. They may write a report. There are chances to educate the people on meaningful subjects. Each animal and/or sacrifice aligns to a person and/or calling to do something.

The animals in the Bible mean tribes or people from certain tribes. They can also mean someone in a certain priesthood/priestesshood position. Here are some of the meanings of the animals in the Law of Moses.

Bull means a presiding Apostle or Disciple.
Calf means a presiding Apostless or Discipless.
Badger means Prince, a presiding Minister (Grand Priest) or Maiden (Grand Priestess).
Badger can also mean a Gadite.
Ram means a presiding Handmaiden or a Steward.
Ram can also mean someone chosen by a presiding Handmaiden.
Shoulder of a ram means someone presiding as a Chieftess Handmaiden.
Lamb means a presiding Servant or Maid.
Goat means a presiding Manservant or Maidservant.
Kid of the goats means the same as goat, except this is someone who isn't in a Chief(tess) position.
Turtledove means someone presiding in a Chif(tess) position.
A young pigeon means someone presiding in a Leader(ess) position.
Ox means Levite.
Dove means a Benjamite or a presiding in Maiden.
Donkey means spouse.

The shekel of the sanctuary means 3.65. This can be used to calculate things such as a number of people or an amount of time.

Identification Keys

There are certain words used to help translate meanings of stories into particular people. This was helpful in deciphering the galactic map in Book 7 of this series.

Grace means Susanna (meaning John the Baptist's wife).
Faith means John the Baptist.
Spear means Gadite – David's seer and/or tribe of Gad.
Hammer means Judas Maccabee of the tribe of Reuben.
Justice means James, son of James Zebedee.
Judgment means James Zebedee.
Mercy means Bathsheba.
Truth means David.
Righteousness means Jesus.
Peace means Mary Magdalene. (Psalm 85:10)

The Galactic Map in Book 7 of this series is a helpful translation device. The animal alignments assist in merging all doctrines.

As for the galactic astrology, it only has constructive words. You may notice that it was completely set into paradise. Everything in that map is beneficial as if there wasn't anything bad existing. It was designed correctly with the Covenant Ark, the Urim and Thummim, and the Inner Oracle. To help align the astrology, the Oracle was used to convert the mythological doctrines of the gods. After conversion, the alignments were cleansed.

Reading the Inner Veil

The inner oracle changes many words into their antonyms. This is designed to change a doctrine that seems like it is about war into something spiritually useful. (Romans 3:1-4) (Hebrews 5:12) (1 Peter 4:11) Solomon revealed an antonym key. Keys like this are what he used to write doctrine. The inner Oracle keys only shift into the good. Love doesn't turn back into hate. (1 John 4:8) (Ecclesiastes 3:1-8)

Die means born.
Pluck means plant.
Kill means heal.
Break down means build up.

Weep means laugh.
Mourn means dance.
Cast away stones means gather stones.
Refrain from embracing means to embrace.
Lose means gain. (Philippians 3:8)
Throw away means keep.
Tear means sew.
Keep silence means to speak.
Hate means love.
War means peace.

There are also Oracle keys that have dual dichotomization.

Beginning and end have dual dichotomization. (Revelation 22:13)
First and last have dual dichotomization. (Matthew 19:30)

The Oracle works with many antonyms. In these next keys, we notice that the Oracle isn't only built in opposites. It also has abstract and metaphor translators.

The inner Oracle doesn't translate good into bad. It shifts everything towards the good. Good and evil don't trade places. The oracle turns evil to good while good remains good. Light and darkness don't swap. Bitter and sweet don't exchange. (Isaiah 5:20) Bitter becomes sweet, yet sweet doesn't become bitter. (Proverbs 27:7) Yes remains yes, while no turns to yes. (2 Corinthians 1:17-19)

Paul speaks of the Oracle, calling it tongues. Long ago the doctrinal tongues of the people were divided. (Psalm 55:9) The Oracle translates the many tongues back into one. (Isaiah 66:18) (Mark 16:17) (Acts 2:1-3, 2:11, 19:6) (1 Corinthians 12:28, 13:1)

(1 Corinthians 14:18-20) "I thank my God I speak with tongues more than you all; yet in the church I would rather speak five words with my understanding, that I may teach others also, than ten thousand words in a tongue. Brethren, do not be children in understanding; however, in malice be babes, but in understanding be mature. In the law it is written: "With *men of* other tongues and other lips I will speak to this people; And yet, for all that, they will not hear Me," says the Lord. Therefore tongues are for a sign, not to those who believe but to unbelievers; but prophesying is not for unbelievers but for those who believe."

Here are several words of the Oracle that Paul revealed.

1: People means field and/or building (GOD's). (1 Corinthians 3:9)
2: Body is Temple. (1 Corinthians 3:16, 6:19)
3: Rod is love (gentleness of Spirit). (1 Corinthians 4:21) Rod is also the Word of GOD.
4: Ox means Levite. (1 Corinthians 9:8-11)
5: Rock means Christ. (1 Corinthians 10:4)
6: Cup is blood(line). (1 Corinthians 10:16)
7: Bread is body.

The many tongues that the people spoke will cease. First their doctrines will go into darkness and die. This happens because they seem to suggest that GOD supports physical war. (Romans 3:13) (1 Corinthians 13:8) (Revelation 7:9, 10:11, 11:9, 16:10, 17:15) The Oracle is designed to gather the doctrines into one and return the clean language of love. This resurrects their words. (1 Corinthians 14:39)

Here is another Oracle key in the book of Revelation. (Revelation 1:20)

Stars means angels. (Nehemiah 9:23)
Lampstands means churches.

Sometimes one word can have a dual translation. For instance, in Genesis, stars means children, while in Revelation it means angels. (Genesis 37:9-10) Stars means angels and children. Cloud means marriage and witnesses (group of). (Acts 1:9) (Hebrews 12:1)

It is important to know that the inner Oracle doesn't always share mirror dichotomization between two words. Many words shift into parallel. This means that yes remains yes while no becomes yes. Some words shift into perpendicular. This means that one word becomes a second word while the second word becomes a third.

Here are some perpendicular translation examples.

Tear means sew, while sew means reap.
Stars means angels, while angels means reapers. (Matthew 13:39)
People means field, while field means world. (Matthew 13:38)
Darkness means light, while light means truth. (Psalm 43:3) (John 3:21)

Here is a list of inner Oracle keys.

Eat means read. (Jeremiah 15:16) (Ezekiel 3:1-2) (Revelation 10:8-10)

Fish means a writing (like a book).

Basket means book. (Jeremiah 24:1-2)

Stomach means mind.

Crown means wife. (Proverbs 12:4)

Blood means resources.

Tribulation means glory. (Romans 5:3) (Ephesians 3:13)

Palm tree means Levite.

Fir tree means Simeonite.

Temple means body. (John 2:21)

Hades means heaven.

City means daughter. (Lamentations 2:15)

Joy means daughter. (Lamentations 2:15)

Hope means son. (Hebrews 3:6)

Tower means son.

Anchor means bloodline. (Hebrews 6:19)

Ox means slave and Levite.

Slave means free. (Romans 6:18-22) (1 Corinthians 7:22)

Camel means servant.

Green means spiritual or filled with Spirit.

Hill means prince.

Mountain means king. (Revelation 17:9-12)

Horns means kings. (Revelation 17:12)

Crooked means straight. (Ecclesiastes 7:13) (Isaiah 40:4, 42:16, 45:2)

Thunder means voice. (2 Samuel 22:14) (Job 37:2-5) (Job 40:9) (Revelation 6:1)

Lightning means power.

Wine means prophesy.

Tent (booth) means wife.

Church means bride. (Ephesians 5:22-24)

Water means spirit.

Seed means word.

Sands means descendants (of Abraham). Like the pyramids were built upon.

Light means truth.

Clothes mean family.

Incense means prayer. (Revelation 5:8)

Censer means prophecy.
Wing means spirit.
Height means man.
Width means woman.
Length means man and woman together.
Death means change. (Job 14:14)
Dust means heaven.
Fragrance means motive.
Jerusalem means Earth.
Locks of hair means experience.
Shave means unveil.
Nation(s) means people(s).
Honey means wisdom.
Christ means teacher when found singly. (Matthew 23:8-10)
Jesus Christ remains the same forever. (Hebrews 13:8)
Liver means soul.
Soul means memory.
Spirit means personality.

Kidneys means mind (lobes of). (New World Translation 1984 Edition Revelation 2:23) "And her children I will kill with deadly plague, so that all the congregations will know that I am he who searches the kidneys and hearts, and I will give to YOU individually according to YOUR deeds."

Unbelief means the world. (Gnostic Interpretation of Knowledge) "But it is a great thing for a man who has faith, since he is not in unbelief, which is the world. Now the world is the place of unfaith and the place of death."

CODEX XI Translated by John D. Turner Selection made from James M. Robinson, ed., The Nag Hammadi Library, revised edition. HarperCollins, San Francisco, 1990.

Body means garment (or group of people).

(Gnostic Sentences of Sextus) "(346) Say with your mind that the body is the garment of your soul: keep it, therefore, pure since it is innocent."

CODEX XII Translated by Frederik Wisse Selection made from James M. Robinson, ed., The Nag Hammadi Library, revised edition. HarperCollins, San Francisco, 1990.

The Four Winds

Bone means prophesy. (Ezekiel 37:1-14)

The book of Ezekiel says that the bones of Israel are dry without hope. It explains that the breath of life comes from the four winds. The four winds (voices) are also known as the four riverheads of doctrine. (Psalm 147:18) Together they resurrect Israel. The dry bones (prophecies) of the Bible become moist again.

There was a story about Abraham in National Geographic Magazine. The reports spoke of how he had traveled in different directions throughout life. These directions were such as north, east, south, and west. Here is how to translate the meaning.

North Wind – Shem – Jews and Christians. (Seals Ham and Japheth) (Proverbs 25:23)
South Wind – Ham – Muslims and Baha'is. (Seals Shem and Oriental Japheth) (Job 37:17)
East Wind – Oriental Japheth – Oriental alternate religions. (Seals Ham and Oriental Japheth) (Exodus 10:19)
West Wind – Japheth – Alternate religions (non-Orient founded). (Seals Shem and Japheth) (Exodus 10:13)

Exaltation of the religions comes from the Word of GOD. That means the Bible. This next verse reveals that Shem is the north wind. (Psalm 75:6) Ham would be opposite Shem in the south. The remaining two winds are for Japheth.

Those are Urim alignments. For the Thummim settings, switch the positions of Shem and Ham.

(Matthew 24:321) "And He will send His angels with a great sound of a trumpet, and they will gather together His elect from the four winds, from one end of heaven to the other."

(The Prophets Surah 21:81) "We harnessed the stormy wind for Solomon, so that it sped by his command to the land we had blessed – We have knowledge of all things." (From the M. A. S. Abdel Haleem edition.)

(Song of Solomon 4:16) "Awake, O north *wind,* And come, O south! Blow upon my garden, *That* its spices may flow out. Let my beloved come to his garden And eat its pleasant fruits.?

The winds are not directly in their atomic positions on the Urim. This is because they are blowing around. Voices of doctrine are meant to be heard. When the angels hold the winds at the four corners of the earth, then they enter their atomic setting. (Revelation 7:1-3) Winds blow in the opposite direction of their position.

Northwest Wind – Atomic – Shem – Held in the southeast.
Southeast Wind – Atomic – Ham – Held in the northwest.
Southwest Wind – Atomic – Japheth – Held in the northeast.
Northeast Wind – Atomic – Oriental Japheth – Held in the southwest. (Acts 27:14)

The four cardinal winds enter the Urim slightly different than in their atomic positions. Voices are carried. They move for reflection and sealing. In their atomic setting, the wind of Oriental Japheth is in the southwest instruction location. That is the prime color yellow area. This truth is signified by the Yellow Emperor of Shenism and the Yellow River. There Japheth relates to the proton in the atomic nucleus. The rest of Japheth enters the northeast understanding location. Wind arrangements are designed to propel the transfer of understanding with instruction.

Living water resurrects the prophecies of the Bible. It can only be produced by uniting the families of Noah and Abraham. Listen to all of their voices (winds). They will bring rain. (1 Kings 18:42-43)

(Gnostic Discourse of the Eighth and the Ninth) "And by a spirit he gives rain upon everyone."

CODEX VI Translated by James Brashler, Peter A. Dirkse, and Douglas M. Parrott Selection made from James M. Robinson, ed., The Nag Hammadi Library, revised edition. HarperCollins, San Francisco, 1990.

Keys of Paradise

Once the winds are anointed, paradise opens. The entire Bible is like a photo negative. Its meanings were hidden. It's an entirely different story behind the veil. (Song of Solomon 4:12-15) Have you ever asked yourself where all the flowers in the Bible went? GOD is love. Where is all the peace and harmony? That is under the veil.

The Oracle in the Bible was written so perfectly that the entire book can turn into paradise. We find a warning at the end of the book of Revelation that we are not to add or take away from the prophecy of the Bible. (Revelation 22:18-19) The reason for that warning is that each word is within a synchronized position. If I change one word in the Bible, such as scales, then that word changes everywhere in the book. Scales means pride. (Acts 9:18) (Job 41:15)

Adding even one sentence to the Bible could damage the Oracle. If you were to place the book in a word program, you could use these Oracle keys. You would change every word into its Oracle translation. If you do this throughout the entire Bible, the flowers begin to bloom. It doesn't change the book; it extracts a photo positive. The Bible remains. Don't add, nor take, from the original text. The Oracle reveals paradise. This process must be done very carefully. Here are more inner veil keys.

Wild changes to tame.
Forget changes to remember.
Rough changes to smooth. (Luke 3:5)
Reject changes to accept.
Decrease changes to increase. (John 3:30)
Yesterday changes to tomorrow.
Take changes to give (offer).
Give changes to receive.
Deaf changes to listening.
Blind changes to seeing.
Dumb changes to speaking.
Lame changes to walking.
Silent changes to speaking. (Acts 18:9)
Sin changes to scarlet. (Isaiah 1:18) Scarlet then becomes white. And white means righteousness.

When completing the inner veil, certain words such as prepositions, definite articles, adverbs, conjunctions, and adjectives can be multi-interchangeable. The way to use these will be found during final translation. Here are several examples.

of/the
now/when
on/in
lay/sit/stand

Names in the Bible reflect the actions of the person. For instance, Midian means strife. Jacob means supplanter. Parents don't usually give their children titles of negativity. This reveals that the names in the Bible have purpose. When entering Paradise, names of disapproval are also shifted to blessed Hebrew names.

When the translation is complete, the Bible becomes a mirror or perfection. It is the story of the Garden of GOD. (Song of Solomon 1:7, 4:1-3, 5:7-8, 6:6-7) (Ezekiel 28:13)

Entering Paradise

The Bible has directions on how to develop its photo negative. These next guidelines are not all direct word shifts. Many of these are principles of the Oracle.

When entering Paradise, the Biblical desert blossoms like a rose. (Isaiah 35:1-2)
The eyes of the blind become opened.
The ears of the deaf are unstopped.
The lame can leap.
The dumb can sing.
The wilderness becomes waters.
The desert is changed into streams.
Parched ground becomes a pool.
The thirsty land becomes springs of water. (Isaiah 35:5-7)
Desolate heights become rivers.
The valleys become fountains.
The wilderness becomes a pool of water.
The dry land becomes springs of water. (Isaiah 41:18)
The wilderness becomes waters.
The desert becomes rivers.
The wilderness becomes Eden.
The desert becomes the garden of the LORD. (Isaiah 51:3)

The Bible was purposely designed broken. There is a man crying in the wilderness. It is to be made straight for the people. Paradise returns. (Isaiah 40:3) (Matthew 3:1-3)

Trees means men (people). (Psalm 1:1-3) (Isaiah 55:12) (Mark 8:22-25) (2 Esdras 4:13, 5:5) In the Torah, Moses counts men for battle. These soldiers mean trees. (Numbers 1:17-44) The 59,300 counted warriors were planted trees. Every soldier should be multiplied by the number of trees that each person is responsible for planting. Currently it is said that Earth is missing 15,000,000,000. Humans have cut them down. Chapter 11 of Book 8 of this series can help calculate tree responsibilities.

What this portion means is that while the Bible is being developed, the Earth is being renewed. (Exodus 34:10) (Psalm 104:30) Mother Earth is aging. Her children must now take care of her.

Non-Paradisical Doctrines

Seed means word. The good seeds of scripture are the children of the kingdom of GOD. (Matthew 13:38) Bad seeds are not of the kingdom of GOD. The Oracle is a great test to see which doctrines are acceptable.

When a doctrine is from GOD, the Oracle can be used to translate it. Consider the doctrines of the laws of the worldly nations. None of their books can be translated with the Oracle. That is how we know that they are not from GOD. They were not written with loving metaphor.

Their doctrines cannot be fixed. (Ecclesiastes 1:15) (Isaiah 50:2) Here is what the LORD says will happen to their doctrines. This includes their forbidden laws and statutes.

(Isaiah 34:8-10) "For *it is* the day of the LORD's vengeance, The year of recompense for the cause of Zion. Its streams shall be turned into pitch, And its dust into brimstone; Its land shall become burning pitch. It shall not be quenched night or day; Its smoke shall ascend forever. From generation to generation it shall lie waste; No one shall pass through it forever and ever."

(Psalm 107:33-38) "He turns rivers into a wilderness, And the watersprings into dry ground; A fruitful land into barrenness, For the wickedness of those who dwell in it.

Men and Women

In the beginning, GOD created mankind. Under the veil, a woman is someone who is not of the LORD. A man is someone who follows GOD. When the Qur'an speaks about women being less than men, woman means worldly governing forces. When the Qur'an says not to become a woman, it means don't merge your religion with the forceful government. It also means not to work for them.

Knowing that woman means someone of the oppressive worldly governments is important. These next verses reveal the symbolization. (Jeremiah 50:37-38, 51:30) (Isaiah 3:12, 9:16, 19:16, 27:11) (Ezekiel 18:6) (Romans 1:24-26) (Nahum 3:1-19) (Jeremiah 31:22)

The Bible speaks against witches. What are they? Witches are those who sew magic charms on their sleeves. They do this to justify hunting the souls of men. These magic charms represent the service patches on the sleeves of officers' clothing. (NKJV Ezekiel 13:17-22) Their ways are witchcraft. They hunt the people and put them into prisons. Prisoners are convicted without ever being allowed to communicate what happened. The entire court system is a system for gaining revenue by selling souls. (Revelation 18:11-13)

Adam and Eve were Mankind. (Genesis 5:2) Eve was tempted and ate from their ways. There the metaphor began. It isn't male chauvinism. We can all be men about this.

(Gnostic The Tripartite Tractate) "their fathers who are the ones who gave them life, each one being a copy of each one of the faces, which are forms of maleness, since they are not from the illness which is femaleness, but are from this one who already has left behind the sickness."

CODEX I Translated by Harold W. Attridge and Dieter Mueller Selection made from James M. Robinson, ed., The Nag Hammadi Library, revised edition. HarperCollins, San Francisco, 1990.

(Gnostic The Hypostasis of the Archons) "Then the female spiritual principle came in the snake"

The Reality of the Rulers CODEX II Translated by Bentley Layton Selection made from James M. Robinson, ed., The Nag Hammadi Library, revised edition. HarperCollins, San Francisco, 1990.

(Gnostic Dialogue of the Savior) "The Lord said, "Whatever is born of truth does not die. Whatever is born of woman dies.""

(Gnostic Dialogue of the Savior) "The Lord said, "Pray in the place where there is no woman." Matthew said, "'Pray in the place where there is no woman,' he tells us, meaning 'Destroy the works of womanhood,' not because there is any other manner of birth, but because they will cease giving birth."

CODEX III Translated by Stephen Emmel Selection made from James M. Robinson, ed., The Nag Hammadi Library, revised edition. HarperCollins, San Francisco, 1990.

(Gnostic Paraphrase of Shem) "He humbled the dark womb in order that she might not reveal other seed from the darkness."

(Gnostic Paraphrase of Shem) "Her likeness appeared in the water in the form of a frightful beast with many faces, which is crooked below."

The Paraphrase of Shem CODEX VII Translated by Frederik Wisse Selection made from James M. Robinson, ed., The Nag Hammadi Library, revised edition. HarperCollins, San Francisco, 1990.

(Gnostic The Second Treatise of the Great Seth) "And do not become female, lest you give birth to evil and (its) brothers: jealousy and division, anger and wrath, fear and a divided heart, and empty, non-existent desire. But I am an ineffable mystery to you."

CODEX VII Translated by Roger A. Bullard and Joseph A. Gibbons Selection made from James M. Robinson, ed., The Nag Hammadi Library, revised edition. HarperCollins, San Francisco, 1990.

(Gnostic The Apocalypse of Peter) "But those of this sort are the workers who will be cast into the outer darkness, away from the sons of light. For neither will they enter, nor do they permit those who are going up to their approval for their release." "And still others of them who suffer think that they will perfect the wisdom of the brotherhood which really exists, which is the spiritual fellowship of those united in communion, through which the wedding of incorruptibility shall be revealed. The kindred race of the sisterhood will appear as an imitation. These are the ones who oppress their brothers, saying to them, "Through this our God has pity, since salvation comes to us through this," not knowing the punishment of those who are made glad by those who have done this thing to the little ones whom they saw, (and) whom they took prisoner."

CODEX VII Translated by James Brashler and Roger A. Bullard Selection made from James M. Robinson, ed., The Nag Hammadi Library, revised edition. HarperCollins, San Francisco, 1990.

Know this fact. We are not here to harm anyone. We are not going to attack the government. They are also just children. What they are doing is very harmful to everyone. Their behavior has become a problem. The officers with guns in their hands are the women, and they are witches.

Without the Oracle, the Bible dies. With the Oracle, the Bible is resurrected. Men lead the way. Let GOD be your guide. (1 Kings 1:52, 2:2) (Sirach 9:2, 25:24, 42:14)

Did GOD Create Everything?

People have asked how GOD created everything yet only created good. If GOD created everything then where did the bad things come from? Whatever GOD created was from love and is only an emanation of love. This has been hard to comprehend. Here is an abstract way of explaining this. GOD created all good acceptable things. GOD didn't create anything bad, and therefore that which GOD didn't create was mad, loathing, hateful, and angry. That which GOD didn't create warred.

(Gnostic Teachings of Silvanus) "But since you cast from yourself God, the holy Father, the true Life, the Spring of Life, therefore you have obtained death as a father and have acquired ignorance as a mother. They have robbed you of the true knowledge. But return, my son, to your first father, God, and Wisdom, your Mother, from whom you came into being from the very first in order that you might fight against all of your enemies, the Powers of the Adversary. Listen, my son, to my advice. Do not be arrogant in opposition to every good opinion, but take for yourself the side of the divinity of reason."

CODEX VII Translated by Malcolm L. Peel and Jan Zandee Selection made from James M. Robinson, ed., The Nag Hammadi Library, revised edition. HarperCollins, San Francisco, 1990.

Touching the Ark of GOD

The Ark of GOD consists of two cherubim, one on each side of a throne. It was taught that if anyone touched the Ark incorrectly, they would die. (Leviticus 16:2)

First, what is the Ark of GOD?

It is a design for the throne of Isaac. It also represents the lineage of Jesus.

How was it built?

We first need a branch of wood to build the Ark with. The branch began as two sticks. One of the meanings of these sticks is two lineages, one from Judah and one from Ephraim. (Ezekiel 37:16-19) They were then put together to form the branch. (Zechariah 3:8-9, 6:9-15) This branch signifies the lineage of Jesus. He is a descendant of Joshua and David. (Zechariah 12:8, 13:1) (Ezekiel 34:23, 37:24-25) The book of Zechariah speaks of these sticks. He explained that the bow is Judah, and the arrow is Ephraim. (Zechariah 9:13) It is the rain-bow. This means that the king of Judah is given the Urim.

To comprehend how the king's Urim is set up, we first study his parents. The parents of the king of Judah were on the throne before him.

While as king and queen, the parents are sealed with the rain-bow. Each of them has eight attendants. These eight people administer in the wisdom, discretion, instruction, counsel, knowledge, perception, understanding, and equity positions of the Urim and Thummim.

The king's Urim has eight female attendants. He sits in a center prudence position. Whoever the king is married to is the queen. If the biological parents split for any reason, the dad remains while the mom's position is offered to another. The queen's Thummim has eight male attendants. She sits in a center prudence position. They are harmless and stand by the LORD's throne, their son.

The attendants of the king and queen are sent out into the world. This could mean local or long distance. They are referred to as horns and eyes. (Revelation 5:6) (Zechariah 3:9) These attendants can be given assignments. They gather information and bring it back to the parents.

The attendants are also known as seven lamps. The eighth of the attendants of each Urim is considered drawn out from the others and therefore isn't mentioned. These lamps (attendants) stand by the two olive trees. The olive trees are the king and queen of Judah as centers of the Urim and Thummim. Here is where the Bible mentions this. (Zechariah 4:1-14)

These two olive trees are thornless and harmless. They are not here to tempt people into wrongdoing. They won't hurt anyone. Their path is to perform GOD's will.

In the center of the Ark, there is a mercy seat. It is considered the throne of GOD. The king and queen are part of the ark. The mercy seat is separated into two parts: truth and mercy. The king is the LORD's throne of truth. The queen is GOD's throne of mercy. The LORD our GOD enters them. (1 Kings 2:4, 3:6) (Psalm 89:14) (Proverbs 20:28, 29:14) (Isaiah 16:5) Though GOD is ONE, we use the words LORD and GOD to explain the different things that GOD does. Otherwise, there isn't any difference.

The king and queen have a network. This network is the crown and throne of the Priesthood. To find the network design, we study the tabernacle of Moses. The Bible's veil allows for one prophecy to have several meanings. In this veil, we begin with the second cover of the tabernacle. It had two sets of curtains. One set had six, and the other had five. (Exodus 26:7) Each curtain signifies one of the kings of the twelve tribes of Israel and his assembly. The curtains are a Urim of one king and eight attendants. They are networked together. These Urim assemblies are what it means by cherubim designs on the curtains. The king of the tribe of Levi doesn't have one.

The curtains (Urim assemblies) are in order pertaining to the division of the land in Ezekiel chapter 48. The order of the six curtains is Dan, Asher, Naphtali, Joseph, Reuben, and Judah. Levi is then separated. The order of the five curtains is Benjamin, Simeon, Zebulun, and Gad. The clasps connecting the two sets of curtains signify the Levites and the division of Abijah. The Levites administer there in the temple center.

A king's Urim assembly has two attendants who network with another king. This weaves them together. Each king's equity attendant is also another king's counsel attendant. This means that although each king has eight attendants, only six are particular to their own Urim. (Isaiah 6:2) (Revelation 4:8) There are eleven kings for eleven curtains. They are connected in this order: Dan, Asher, Naphtali, Joseph, Reuben, Judah, Benjamin, Simeon, Zebulun, Gad, and back to Dan. The king of Judah is the Preacher. The next page has a diagram of the network.

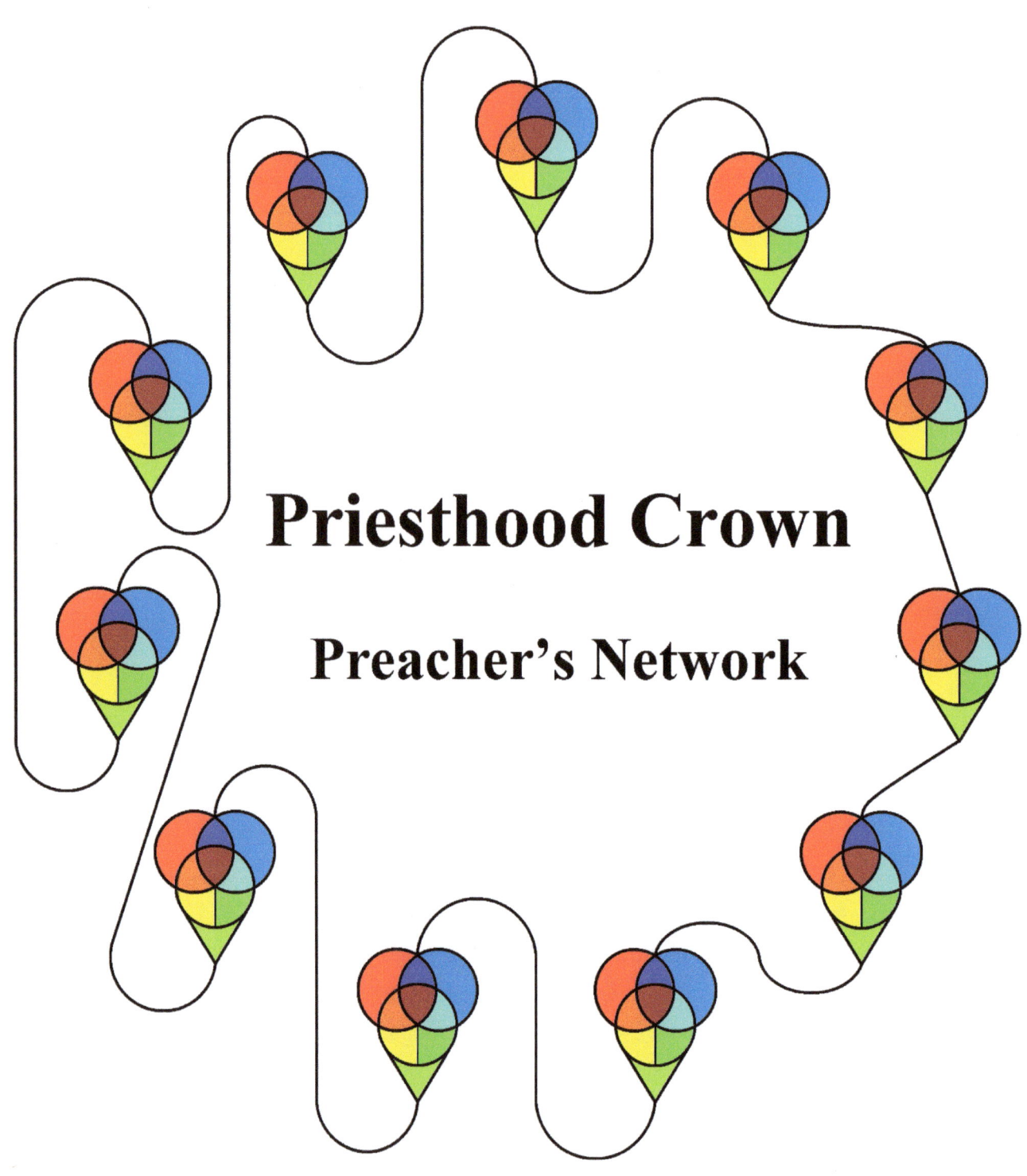

In the design of Moses' tabernacle, tent refers to the kings because this priesthood connects them as one with the queens. Otherwise, tent means wife. The queens' Thummim assemblies are connected just like the kings' curtains. The kings are networked with female attendants. The queens are networked with male attendants. These attendants are sent out into the world with assignments. They gather information for their assembly. The assemblies have cross collaboration through the counsel and equity positions.

The king of Judah's eleventh curtain doubles over the back of the tabernacle. That is for reciprocation between kings and queens. (Exodus 26:9) There is a second set of ten curtains. These ten are the queens' network. (Exodus 26:1) Mentioning only ten means that the eleventh assembly of the queen of Judah reports to the king of Judah. That is where they become one. The queens gather information through their attendants. They make decisions together. The assembly of the queen of Judah then delivers the queen's report to the King of Judah. The king of Judah then delivers the report to the kings' assembly. That is the hidden tent key. If it were the other way around, we would have another beast and harlot design. (Revelation 17:12-18)

The network is first aligned with the tabernacle and millennial temple grounds. It is then finalized with Solomon's temple. The Preacher's temple had a network of 200 pomegranates. This means that the assemblies of kings and queens must have 100 people each. The king and queen of Levi are included. The Archbishop of Levi (Aaron) and his wife are included. The Evangelist (John the Baptist) of the division of Abijah (sons of Zadok) and his wife are included. The Prophet (Jacob) and his wife each have a free standing Urim and Thummim. The Prophet and Prophetess don't have mixed gender assemblies.

Here is the complete networking.

Kings Assembly
Kings' Urims (Curtains) 11x8 = 88 (eleven men and 77 women)
King of Levi – 1
Aaron – 1
Abijah Evangelist (John the Baptist) – 1
Prophet's (Jacob's) Urim – 9 men (the badger skins) (Exodus 26:14)

Queens Assembly
Queens' Thummims (Curtains) 11x8 = 88 (eleven women and 77 men)
Wife of Levi – 1
Aaron's wife (Elisheba) – 1
Abijah Bishop's wife – 1
Prophet's (Jacob's) wife's Thummim – 9 women (the badger skins) (Exodus 26:14)

Each network side, kings and queens', has 100 people.

(Jeremiah 52:23) There were ninety-six pomegranates on the sides; the total number of pomegranates above the surrounding network was a hundred.

Together the networks make 200 people.

(1 Kings 17:20) "On the capitals of both pillars, above the bowl-shaped part next to the network, were the two hundred pomegranates in rows all around."

A man is chosen to be king when he is at least thirty years of age. He becomes the practicing king at thirty-four years or later. A king of Judah may retire after twenty-one years of being king. That is three sabbatical years. They then blow the trumpets for the new king. When he retires, so does the entire assembly of 200 people. A son of King David becomes the next king of Judah. (Psalm 132:11) The next king begins a completely new assembly of 200 people. His parents then enter the position of the cherubim on the sides of the throne. They are symbolized by the red ram skins on the tabernacle. (Exodus 26:14) The previous 200 people remain with the retired king and queen. They can continue in kingdom needs. The previous network must remain very close for at least five years. After five years, they can enter Evangelist lives if they choose. They remain in their positions until the second king after them. They are there to help. The previous network can deliver reports to the two cherubim by the throne. The two cherubim (new king's parents' Urim assemblies) remain in open contact with the new king and queen. It is a thirty-year job, including the transition schedules. There is much more information on this in books 6 and 11 of this series.

These two kings' networks together are the 400 people double network.

(2 Chronicles 4:13) "So Huram finished the work he had undertaken for King Solomon in the temple of God: the two pillars; the two bowl-shaped capitals on top of the pillars; the two sets of network decorating the two bowl-shaped capitals on top of the pillars; the four hundred pomegranates for the two sets of network (two rows of pomegranates for each network, decorating the bowl-shaped capitals on top of the pillars);"

The dad of the king of Judah holds the Abraham position. The king of Judah holds the Isaac position. The Prophet, also known as David's seer, holds the Jacob position. (Matthew 8:11)

The King, the parent's son, is as the throne of the LORD. He isn't GOD; rather, the LORD sets HIS will upon him. HE speaks through the king.

Administering Angels

The King and Queen of Levi are not connected to a Urim and Thummim. Levites are considered the LORD's personal possession. (Numbers 3:12) Their Levitical Priesthood has the keys for the administration of angels. The kings and queens are networked together to receive that blessing.

Here is the order of angels that administer to the network.

One Seraph is comprised of nine angels. One Cherub and eight Ophanim make a single Seraph. Each Cherub has eight Ophanim called wings. They are sealed like a Urim.

One Cherub administers to each person holding a center position on a Urim. One Ophanim administers to each attendant of the Urim. It is the same for a lady's Thummim. There are twenty-three Urims and twenty-three Thummims in the 400 people network. Forty-eight seraphim administer to the crown priesthood.

The Cherubim administering to the kings and queens each have Ophanim. They are sometimes called Thrones or wings. Two wings of each Cherubim administering a king or queen is also the Ophanim of another Cherub. This means that each Cherub of the throne has six exclusive wings (Ophanim) to their Seraph administration. (Isaiah 6:1-9) These angels are not worshiped at all. They are only helpers.

Touching the Ark

Why would handling the Ark of GOD in an unapproved way be harmful?

Adam, Solomon, and Jesus were each called the Son of GOD. (Luke 3:38) (1 Chronicles 28:6) The Son of GOD was a specific lineage that passed through King David. (Revelation 22:16) The Ark is the lineage of Jesus. It has the king and queen of Judah in the center. There the LORD GOD has them as a throne to set upon. They do GOD's will. They are a private possession. When the king and queen of Judah retire, they become the cherubim beside the throne.

These people are the seal of the LORD's authority here on Earth. They don't have the power to harm others. They cannot save people by force. They cannot physically break people out of prisons or go to war. They cannot destroy others. They have the ability to do normal things here. And they have GOD's guidance.

To threaten or harm them is a direct attack against GOD. Losing them would result in the extinction of humanity. These people can seem slightly alien to others. Because of this, the LORD protects them.

It is said that if the Ark is captured or touched incorrectly, then people die. (1 Samuel 4:11-18) (2 Samuel 6:7) (1 Chronicles 13:10) What this means is that if someone attempts to touch these people by force, the LORD intervenes. We don't always know how GOD will react.

In these next verses, someone dies for touching the Ark. (2 Samuel 6:6-7) The oxen (Levites) have stumbled, meaning that they bowed to the government. Because the Levites stumbled, they merged their church into carnal laws. This is called sexual immorality. They mated with the Devil. (Revelation 2:14) Once they mated with the Devil, people then thought it was ok to use carnal laws against GOD.

The parents and their son, the king of Judah, are the Ark of GOD. It is illegal to touch the Ark with an officer of the worldly laws. So far as it is known, at least three people have attempted to correct (stabilize) the Ark. They did so by involving police officers. All three of those people lost their firstborn child. The LORD simply dropped their firstborn children for touching HIS child. When anyone turns to an officer to solve their problems, it means they turn to Egypt for protection. Leviathan means government, and it also means Pharaoh. (Ezekiel 29:3, 32:2) When those three parents chose to remain with their governing forces instead of the LORD, their three children were counted with the firstborn of Egypt. Information about this with scientific proof is in Book 5 of this series.

(Barnabas 6:1-2) "When then He gave the commandment, what saith He? *Who is he that disputeth with Me? Let him oppose Me. Or who is he that goeth to law with Me? Let him draw nigh unto the servant of the Lord" Woe unto you, for ye all shall wax old as a garment, and the moth shall consume you.* And again the prophet saith, seeing that as a hard stone He was ordained for crushing; *Behold I will put into the fountains of Zion a stone very precious, elect, a chief corner-stone, honorable.*"

Because of the life-threatening significance of this fact, the government was notified not to touch these people. They were warned many times. The officers were incapable of caring. They didn't believe. Their only response was to overpower without reason. Due to their actions, at least three children are now lost. The officers are held responsible for their parent's losses. Only three of these incidents have been reported as of the end of 2024. There could be hundreds or even thousands more. An officer is never allowed to touch the Ark. They live outside of the veil, and therefore their penalty is bound to a fallen law.

Touching the Ark of GOD with kindness is permitted.

Temple Body

Jesus stated that His temple is His body. (John 2:21) The tabernacle of Moses was the first temple of GOD. Solomon built the second one. This third one includes the first and second. When the entire body is knit together, that is Christ's body of people. (1 Corinthians 6:15-17) This doctrine is stably founded. The winds of doctrine don't toss it to and fro. Remain clear of carnal doctrines.

(Ephesians 4:15-16) "And He Himself gave some *to be* apostles, some prophets, some evangelists, and some pastors and teachers, for the equipping of the saints for the work of ministry, for the edifying of the body of Christ, till we all come to the unity of the faith and of the knowledge of the Son of God, to a perfect man, to the measure of the stature of the fullness of Christ; that we should no longer be children, tossed to and fro and carried about with every wind of doctrine, by the trickery of men, in the cunning craftiness of deceitful plotting, but, speaking the truth in love, may grow up in all things into Him who is the head—Christ—from whom the whole body, joined and knit together by what every joint supplies, according to the effective working by which every part does its share, causes growth of the body for the edifying of itself in love."

Within the king's network, the Ark of GOD extends to hundreds of people. These people can be anywhere. If an officer touches them with any force, they will have to pay the recompense. The officers are held guilty.

Rebuilding the Temple

Tear this temple down, and in three days I will build it up. (John 2:19-22) When Jesus said that, He was speaking about His body of people. Three days means three thousand years. (2 Peter 3:8) To figure out the correct timing, we go back to when the last real temple was here. It was the temple of Solomon.

(2 Esdras 10:44-48) "The woman told you about the death of her son. Then she vanished from your sight, and a whole city appeared. This is the meaning: The woman you saw is Jerusalem, which you now see as a completed city. When she told you that for thirty years she had had no children, it meant that for three thousand years no offerings had yet been made there. Then Solomon built the city, and sacrifices began to be offered there. At that time the childless woman gave birth to her son. When she told you that she took great care in bringing the son up, that referred to the period when Jerusalem was inhabited. When she told you of the death of her son on his wedding day, that meant the destruction of Jerusalem.

Solomon was born around 1010 B.C. He became king at about twenty years of age. The scientifically proven beginning date for the new era is 12-22-2011. It has been 3,000 years. GOD is now building the millennial temple body. This temple body is the network of 400 people.

Bringing the Ark in

A network like this could be set up by anyone. That would be a body of people. Where would the Ark be, though? The Ark of GOD is the lineage of the King of Judah. When the real Ark is brought into the temple body, GOD's authority enters. It isn't GOD's temple without the Ark.

The LORD said that Solomon was HIS son. He was the Ark of GOD.

(1 Chronicles 17:12-14) "He shall build Me a house, and I will establish his throne forever. I will be his Father, and he shall be My son; and I will not take My mercy away from him, as I took *it* from *him* who was before you. And I will establish him in My house and in My kingdom forever; and his throne shall be established forever.""

(1 Chronicles 22:10) "He shall build a house for My name, and he shall be My son, and I *will be* his Father; and I will establish the throne of his kingdom over Israel forever.'"

(1 Chronicles 28:6) "Now He said to me, 'It is your son Solomon *who* shall build My house and My courts; for I have chosen him *to be* My son, and I will be his Father."

After Solomon, the temple (body of GOD's people) was dismantled. Here is how it was torn down. First, Pharaoh decided who the king would be. Then Nebuchadnezzar decided who the king would be.

(2 Kings 23:34) "Then Pharaoh Necho made Eliakim the son of Josiah king in place of his father Josiah, and changed his name to Jehoiakim."

(2 Kings 24:17) "Then the king of Babylon made Mattaniah, *Jehoiachin's* uncle, king in his place, and changed his name to Zedekiah."

That was the very end of Jerusalem. Any time that any religion allows the government to control their decisions instead of the LORD, they immediately lose GOD. If a church is allowing carnal officers to administer in their positions, they lose their rights. They lose all priesthood authority as soon as they allow a carnal officer to lead their people. This includes police, military officers, veterans, politicians, those who administer their laws, lawyers, judges, practicing psychologists, etc.

When Jesus came, the Romans were in charge of Jerusalem. Even today the people who lead the lands of Jerusalem have joined the United Nations. They are not with GOD. And that is the root of all their problems.

Jesus had already been dead for 2,000 years when the Gospels were written. Jesus is the Word of GOD. Jesus is the lineage of King David. Jesus is a person that came here 2,000 years ago. He was the direct lineage. He was the Ark of GOD. As soon as He was denied, Jerusalem was torn down along with its temple. That one wasn't Jesus' temple; Solomon's was. Today GOD is addressing the world openly instead of only in Jerusalem. Remember what happened to Israel.

A New Kingdom

The kingdom is being given to the people. It is here to save humanity and the Earth. It is designed to remove oppression. Many great gifts are offered. Here are some of them.

1: Great jobs for everyone. Each job pays the same. Pay is allotted as an allowance of resources.

2: Resources include homes, vehicles, food, etc. Each person or family is allowed to have one nice home.

3: The goal is thirty-two-hour work weeks. That is eight hours per day and four days per week. They then have two days to work on their own lives each week. This is accomplished by eliminating unnecessary jobs. Unnecessary, for example, can mean things such as many of the fast-food restaurants. Each job then has more workers, which allows shared time off.

4: If anyone wants to increase their resources, they can work more. They can even do fifty or more hours per week if they willingly choose.

5: Workers are allotted an ample amount of time for vacations. Everyone is allowed to travel the world. Global trips are provided freely and distributed equally.

6: The retirement age is sixty years. They can continue working or become Evangelists if they choose.

7: Free college for everyone who wants it.

8: Full disaster relief is provided. That is one of the main jobs of the kings' network. They are not like monarchs over everyone. The King of Judah has the authority to make great decisions during emergencies. The kings generally allow the people to remain in charge together.

9: The finest fully covered medical and dental for everyone in the kingdom.

Priesthood Connections

The priesthood of Shem is designed to unite all churches. In the city of Zion, there is a chief priesthood group named after Zadok. That is where the priesthood crown and apostles assemble. There is another leading priesthood group in the city of Shiloh. It is connected to the priesthood of Zadok. Each church oversees a portion of the chief and leading priesthood groups. Together these great priesthoods have a little over 1,200 people. (Ezekiel 43:12) When these two groups hold onto the king's network, they are protected by its covenant. As the body of people, they become part of the Ark of GOD.

These chief and leading priesthood groups are known as the train of the LORD's robe. They are connected to the crown priesthood of the kings. (Isaiah 6:1-2) One man from each tribe is called to be a king. Their wives are the queens. It is a network for bringing all churches together.

Each church also receives a priesthood. This gives everyone a position in the kingdom. When the churches with their priesthoods hold onto the train of the LORD's robe, they can be protected by its covenant. This is offered based on their obedience to GOD. As Christ's body of people, they become part of the Ark of GOD.

We are not to seek personal monetary gain over others. Though we aren't paid more in resources depending on job choice, some occupations offer more priesthood authority.

Ham and Japheth are their own families. The three families are not to oppress each other or fight for power. The kingdom doesn't have three heads. The King of Judah is the LORD's throne. In an emergency, all families are to listen to HIM. The kingdom is not to fall into the old ways of the eagle, explained in book eight of this series.

When the families of Ham and Japheth get their keys and hold onto Shem, they can be protected by its covenant. This is offered based on their obedience to GOD. As the body of the LORD's people, they become part of the Ark of GOD.

This is an incomplete illustration. Much more information on the priesthood can be found in books 6 and 11 of this series.

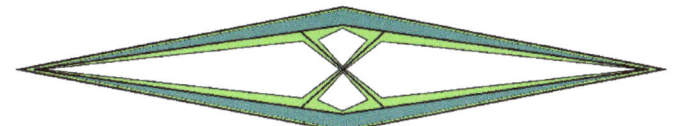

Book of Life Afterward and Epilogue
The Twelfth Sinew of Adamant

Draw near to the Covenant Ark. Fear not. You will die when you touch it. Your death will become a renewing process within your heart and mind. A new person you will become.

The Ark is like a picnic basket. It is set to feed the people. Go ahead, open it up.

Table of the Levitical Church of Phillipia

When you open the Covenant Ark, you would find several things. These include a pot with manna, the Ten Commandments, and Aaron's rod that budded. (Hebrews 9:4)

Manna

Manna has two main meanings. The first meaning is veganism. The second meaning is unveiled doctrine, known as the bread from heaven. Both are needed. Together they are fruits of the tree of life.

The Israelites were supplied with a vegan diet called manna. This assured that they didn't go apostate. Eating meat would kill their ability to hear and obey GOD. We can see the evidence in the story of the quail. Those who complained and yielded to the craving for meat were plagued. (Numbers 11:4-6, 11:31-35)

There is a story in the Talmud about men speaking of apostasy. They are saying that apostasy is connected to notches in a knife. They were explaining it in a way that wouldn't have made sense. Most of them had forgotten. The notches represent how many animals the knife has killed. Killing animals for food leads to apostasy. Veganism supports spirituality.

The bread from heaven is the unveiled meanings of the scriptures. It is sometimes called Jesus' flesh. (John 6:41-58) We eat the Word of GOD. Jesus' flesh is the unveiled Word. It brings life to the doctrines. Without the unveiling, people die. One may be vegan yet still atheist. That death is that they fall away from GOD.

The manna was given so that the people know not to live by bread (biblical word) alone. They are to live by every word that proceeds from GOD. Jesus brought the three-family foundation of the Bible (bread from heaven). The bread of heaven unlocks all of GOD's doctrines. (Deuteronomy 8:3)

Here is some bread from heaven.

The manna was like white coriander seed. This means small seeds of righteousness. It tasted like wafers made from honey, meaning wisdom. It was the color of bdellium, which can be orange, dark red, or white. Orange would mean discretion. Dark red would mean prudence. White would mean prudence and/or righteousness. The manna was gathered, ground up, and cooked in a pan to make cakes. It also tasted like a pastry made with oil. (Numbers 11:7-9)

Here is what that means. The manna was information like small seeds of righteousness. It was broken down with the oracle keys. It was then designed and prepared as the word of GOD, unveiled to be offered as a clean education. It was made sweet like a kindness anointed by GOD. It presented wise discretion for the prudence of each person. They continued eating (reading) this unveiled doctrine of wisdom until they got to the border of Canaan. (Exodus 16:35) When they got to Canaan, they put a veil over the doctrine. This was done so that they could enter without being seen. Canaanites were a military-minded people. They wouldn't accept a truly peaceful doctrine. Peacefulness wouldn't feed the honor of the battle, which supplied Canaan. That is why the Bible was made to look as if GOD encourages war. Canaanites worshiped a god of war, and the Hebrews needed spiritual camouflage.

Moses hid the face of his doctrine behind a veil. (Exodus 34:33-35) (2 Corinthians 3:12-18) He hid the meaning so well that even the Israelites didn't understand it. Without the unveiled Word being unveiled, people died. The unveiled Law of Moses caused the people to believe in war. That way is fallen.

There are two main ways to die. A literal death and a change of mind and heart. The LORD wouldn't allow them to enter paradise until they would change. Otherwise, they could wait outside the gates until the day they die.

(Joshua 5:3-6) "So Joshua made flint knives for himself, and circumcised the sons of Israel at the hill of the foreskins. And this *is* the reason why Joshua circumcised them: All the people who came out of Egypt *who were* males, all the men of war, had died in the wilderness on the way, after they had come out of Egypt. For all the people who came out had been circumcised, but all the people born in the wilderness, on the way as they came out of Egypt, had not been circumcised. For the children of Israel walked forty years in the wilderness, till all the people *who were* men of war, who came out of Egypt, were consumed, because they did not obey the voice of the LORD—to whom the LORD swore that He would not show them the land which the LORD had sworn to their fathers that He would give us, "a land flowing with milk and honey.""

What is circumcision? It means to be detached from the carnal ways of forceful governments. Chapter 4 of this book reveals that we are on the eighth day of creation. (Genesis 17:9-14, 21:4) The New Testament reveals the differences in forms of law. There is GOD's Law and carnal laws, which are not real laws at all. Paul explains that literal circumcision as it was understood isn't the real one. In these next verses, Paul is speaking about two different laws and two different circumcisions.

(Romans 2:25-29) "For circumcision is indeed profitable if you keep the law; but if you are a breaker of the law, your circumcision has become uncircumcision. Therefore, if an uncircumcised man keeps the righteous requirements of the law, will not his uncircumcision be counted as circumcision? And will not the physically uncircumcised, if he fulfills the law, judge you who, *even* with *your* written *code* and circumcision, *are* a transgressor of the law? For he is not a Jew who *is one* outwardly, nor *is* circumcision that which *is* outward in the flesh; but *he is* a Jew who *is one* inwardly; and circumcision *is that* of the heart, in the Spirit, not in the letter; whose praise *is* not from men but from God."

(Romans 3:1-4) "What advantage then has the Jew, or what *is* the profit of circumcision? Much in every way! Chiefly because to them were committed the oracles of God. For what if some did not believe? Will their unbelief make the faithfulness of God without effect? Certainly not! Indeed, let God be true but every man a liar. As it is written: "That You may be justified in Your words, And may overcome when You are judged.""

Keeping the Law of GOD by not joining in the governments of the world is the real circumcision. Dropping the false carnal laws is what Paul means for us to do. Notice that he is speaking of two circumcisions and two laws in one set of verses.

(Romans 3:27-31) "Where *is* boasting then? It is excluded. By what law? Of works? No, but by the law of faith. Therefore we conclude that a man is justified by faith apart from the deeds of the law. Or *is He* the God of the Jews only? *Is He* not also the God of the Gentiles? Yes, of the Gentiles also, since *there is* one God who will justify the circumcised by faith and the uncircumcised through faith. Do we then make void the law through faith? Certainly not! On the contrary, we establish the law."

We learn that through faith, the people of the churches are justified in having been who they were. They had faith in GOD. Even so, they must now become circumcised or their faith becomes futile. Circumcising them is what brings Christ back to life. If Jesus, the Word of GOD, didn't resurrect in them, then they remain dead. (1 Corinthians 15:17) Third Commandment: Thou shalt not take the name of the LORD your GOD in vain. This means that those who claim to be with GOD, if true, would also obey the LORD. (Isaiah 29:13)

(Romans 4:9-12) *Does* this blessedness then *come* upon the circumcised *only,* or upon the uncircumcised also? For we say that faith was accounted to Abraham for righteousness. How then was it accounted? While he was circumcised, or uncircumcised? Not while circumcised, but while uncircumcised. And he received the sign of circumcision, a seal of the righteousness of the faith which *he had while still* uncircumcised, that he might be the father of all those who believe, though they are uncircumcised, that righteousness might be imputed to them also, and the father of circumcision to those who not only *are* of the circumcision, but who also walk in the steps of the faith which our father Abraham *had while still* uncircumcised.

This next set of verses omits the literal circumcision.

(1 Corinthians 7:17-19) But as God has distributed to each one, as the Lord has called each one, so let him walk. And so I ordain in all the churches. Was anyone called while circumcised? Let him not become uncircumcised. Was anyone called while uncircumcised? Let him not be circumcised. Circumcision is nothing and uncircumcision is nothing, but keeping the commandments of God *is what matters.*

Every religion is to be circumcised in the Spirit. (Romans 15:7-13)

Eating with the LORD

First Commandment: There is only one GOD. The governments that don't listen to the LORD are not your GOD. Sixth Commandment: Thou shalt not murder. It doesn't just mean people. That refers to many things, including animals. This brings us to vegetarianism.

The 7th Day Adventists practice vegetarianism and often veganism. We learn from them that the Law requires us all to be vegetarian as a minimum. We know this because the 7th Day Adventists are the Levites who got that Law directly from GOD. (Numbers 3:12, 3:45, 8:14) We find proof of this in the book called The Testaments of the Twelve Patriarchs. The table of the LORD was given to those who would be in the temple on the true Sabbath as in the Fourth Commandment. We are to eat from the LORD's table, which means vegetarian as a minimum requirement. (Isaiah 21:5)

(Testament of Levi 8:16) "Therefore, every desirable thing in Israel shall be for thee and for thy seed, And ye shall eat everything fair to look upon, And the table of the Lord shall thy seed apportion."

(Testament of Judah 21:5) "dominated by the earthly kingdom. For the angel of the Lord said unto me: The Lord chose him rather than thee, to draw near to Him, and to eat of His table and to offer Him the first-fruits of the choice things of the sons of Israel; but thou shalt be king of Jacob."

From The Apocrypha and Pseudepigrapha of the Old Testament by R. H. Charles, vol. II, Oxford Press

Vegetarianism is accepted based on the idea that the animals are our friends. Animals in slaughterhouses are not our friends. They are prey. People rob them of their milk and kill them for their meat. Friends, on the other hand, do share. That's the reasoning of why vegetarian was ok if your animal willingly shared with you.

Eighth Commandment: Thou shalt not steal. This brings us to veganism. We are not to steal milk and honey from the animals.

Aaron had an almond rod that budded. Aaron was the Levite Archbishop. It had been said that GOD had prepared a land flowing with milk and honey. (Exodus 3:8) (Deuteronomy 26:15) Milk means knowledge, and honey means wisdom. (Exodus 3:8) (Jeremiah 32:22) There is another meaning to this rod. It means almond milk, or veganism.

(Jeremiah 1:11-12) "Moreover the word of the LORD came to me, saying, "Jeremiah, what do you see?" And I said, "I see a branch of an almond tree." Then the LORD said to me, "You have seen well, for I am ready to perform My word.""

Bowl of the Reubenite Church of Ephesus

In the beginning, the LORD made Adam from the dust of the earth. HE commanded man, saying that he could eat from every tree except from the tree of the knowledge of good and evil. From Adam, HE made Eve. After they were created, they disobeyed and ate from the tree, which they were not supposed to. The LORD then told Adam that he would return to the dust of the Earth. That was where he came from, and that was where he was headed.

To find out what the forbidden fruit was, we must learn why Adam was the dust of the Earth. What are the characteristics of the Earth? Material Earth has no compassion. The earth isn't considerate of who lives on fault lines during an earthquake. When there is a landslide, the dust isn't compassionate for those killed. Now realize what happened when Adam ate meat for the first time. He was no longer considerate of the animals. He could kill and not care. Like the Earth, he lost compassion for life. He became like the dust of the earth. When people lose compassion for life other than their own, the entire world begins to die. Look at humanity as a species. People waste fuel just to look nice in bigger, fancier vehicles. They cut the forests down without replanting. They use resources without caring to recycle. They haven't been truly considerate. What about the futures of others? What about human-caused extinctions? The path of meat-eating and losing genuine care for everyone will cause the end of humanity. That path kills the earth.

GOD gives us direction to be vegan. We are to do it. The second commandment states that there are not to be any idols (carved images). We are not to bow to them. There are people in charge by GOD's own will. GOD is responsible for them. Those in charge are not to stand above GOD's Word. Many nutritionists with PhDs attempt to use their honors as justification for eating meat and dairy. People have been known to bow to the carved images of their educations.

Many are incorrect, and others are lying. The eat right for your blood type diet is an example of a professional excuse. Those recommending it were leading people incorrectly. Even a cat, who seems to have been designed to only eat meat, is successful as a vegan. Humans can properly feed felines. GOD wouldn't have commanded us not to kill or steal if a blood type prevented us from being vegan. This all depends on what you are putting in your body. People cannot carve any image above this guidance from GOD.

This second commandment also brings us to comprehend the officers. They have carved their images of honor and authority by force. We are not to bow to them. The Priesthood has positions offered by GOD. Those hold the true authority that we can trust.

Bowl of the Danite Church of Smyrna

The first of GOD's ways, before the eating of the forbidden fruit, was veganism. The behemoth ate grass. Veganism is the first step towards the ways of GOD.

(Job 40:15-19) ""Look now at the behemoth, which I made *along* with you; He eats grass like an ox. See now, his strength *is* in his hips, And his power *is* in his stomach muscles. He moves his tail like a cedar; The sinews of his thighs are tightly knit. His bones *are like* beams of bronze, His ribs like bars of iron. He *is* the first of the ways of God;"

Certain human ancestors were once vegan. We can only now see the scientific evidence of the transition to meat-eating. After eating meat, wars between tribes and killing began. There is data about this on these next sites mentioned.

https://www.inverse.com/article/31625-human-evolution-meat-eating-vegan-africa-grassland-hunting

https://www.discovermagazine.com/the-sciences/not-all-prehistoric-humans-loved-meat-some-were-vegetarians

https://www.veganeasy.org/discover/news/new-study-debunks-paleo-diet-reveals-early-humans-ate-mostly-plants/

GOD had once separated a group of humanoids from the ways of the dust. After eating meat, they fell. Without the compassion and consideration, the children of Adam were once again like the dust of the Earth. (Deuteronomy 32:8)

On the other hand, Leviathan, spoken of in Job chapter 40, is the meat eater. It is warlike and loves the battle. Because of eating animals, the murder began. This battle of killing one another won't ever be won. War will forever exist in meat eaters. They love it. Fighting and killing is a sport to them.

(Baruch 3:17) "those who made sport of the birds of the air, and who hoarded up silver and gold in which people trust, and there is no end to their getting;"

Have you ever seen what an orca or a dog does with its kill? They play. They may rub themselves on it and throw it up into the air playing catch. They get so happy when they kill something. That's what Canaan was doing with Jesus for about 2,000 years. They killed the Word of GOD, and had so much fun playing with Him. They even wore Him like leather from a kill. They showed off the honor of their battle against GOD.

Bowl of the Josephite Church of Pergamos

It is wonderful that GOD offers redemption through the Lamb.

All that we eat shapes us. In the Bible, to eat the Lamb first means to read the Word of GOD.

(Jeremiah 15:16) "Your words were found, and I ate them, And Your word was to me the joy and rejoicing of my heart; For I am called by Your name, O LORD God of hosts."

(Ezekiel 3:3) "And He said to me, "Son of man, feed your belly, and fill your stomach with this scroll that I give you." So I ate, and it was in my mouth like honey in sweetness."

(Revelation 10:9) "So I went to the angel and said to him, "Give me the little book." And he said to me, "Take and eat it; and it will make your stomach bitter, but it will be as sweet as honey in your mouth.""

Let us comprehend this Word from GOD.

The Livet, spoken of in Book 8 of this series, doesn't include figs. Many fig trees are basically meat eaters like a Venus flytrap. (Jeremiah 24:3) Fig trees eat wasps to pollinate. When Adam and Eve originally ate from the forbidden fruit, they covered up with fig leaves. To cover up that way means to hide the truth with the reasoning of a meat eater. Adam and Eve hid from GOD by using the reasoning (or lack thereof) of a killer.

To become like a killer means to fall into death. And they were ashamed.

The churches promote the eating of meat. They also have had a shame-based system.

(Gnostic The Teachings of Sylvanus) "For a foolish man usually puts on folly like a robe, and like a garment of sorrow, he puts on shame. And he crowns himself with ignorance, and takes his seat upon a throne of nescience. For while he is without reason, he leads only himself astray, for he is guided by ignorance."

When the people eat meat, they become spiritually dead, buried in the cravings of their natural senses. (Numbers 11:31-34)

(Gnostic The Teachings of Sylvanus) "It is not good for any man to fall into death. For a soul which has been found in death will be without reason. For it is better not to live than to acquire an animal's life."

(Gnostic The Teachings of Sylvanus) "He does not delight in acquiring the light of Christ, which is reason."

(Gnostic The Teachings of Sylvanus) "Do not tire of knocking on the door of reason, and do not cease walking in the way of Christ."

(Gnostic The Teachings of Sylvanus) "Entrust yourself to reason and remove yourself from animalism."

CODEX VII Translated by Malcolm L. Peel and Jan Zandee Selection made from James M. Robinson, ed., The Nag Hammadi Library, revised edition. HarperCollins, San Francisco, 1990.

When the people killed animals, they acted like beasts. The animals became their instructor. The beast became their god. That is why they were cast out of the presence of the LORD. They were cast out by their own wills. Their knowledge of good and evil became animalistic.

(Gnostic On the Origin of the World) "The interpretation of "the beast" is "the instructor"."

"The Untitled Text" CODEX XIII Translated by Hans-Gebhard Bethge and Bentley Layton Selection made from James M. Robinson, ed., The Nag Hammadi Library, revised edition. HarperCollins, San Francisco, 1990.

For millennia, Christians had denied science. They believed that the Earth was the center of the universe. They taught that there couldn't be life on other planets. They saw that the Bible says that Adam was the first man. Because of that, they had denied the science of evolution. As the Gnostics state, they had rejected reasoning. Those who don't kill or steal are a light. The light of Christ is found in veganism. While not eating meat like a beast, veils open. When people hide behind the lack of reasoning of a meat eater, truth gets denied.

Consider an officer. When the police are called, they barge in like brute beasts. They don't listen. Their mandates rule without reason. In North American courts, people are usually sentenced without ever being able to communicate. Prisoners are separated from family. They are then forced to be silent in court during the entire sentencing process. They are only allowed to answer questions that hinder them from explaining. An officer has a meat eater's lack of reasoning. They had taught churches to be like them.

Ninth Commandment: Thou shalt not bear false witness against thy neighbor. The religious leaders didn't know that many of them were from GOD. Churches denied each other. They denied the diversity of Abraham's religions. They were rejecting the doctrinal body of Christ. People were falsely called heathens so that Christians could justify killing them. And not just Christians.

Bowl of the Benjamite Church of Thyatira

When people eat meat the way Adam and Eve did, they hide from the LORD. When people of the churches are asked about the facts of righteousness and being vegan, they hide behind Biblical verses. The Bible was written with a veil. The LORD mirrored their behavior and also hid from them. (Deuteronomy 31:17-18, 32:20) (Job 13:24, 34.29) (Psalm 13:1, 27:9, 44:24, 69:17, 102:2, 104:29, 143:7) (Isaiah 8:17) (Micah 3:4)

The LORD doesn't always mirror. If you lie, the LORD doesn't lie to mirror your behavior. People mirror people. GOD may make something look like a lie, yet the LORD doesn't exactly mirror us. Those who are hiding from GOD live outside of the veil. They believe in and practice from outside the City of GOD. (Revelation 14:20) The Bible mirrors their behavior. They see GOD as a master who guides troops into lands to murder entire populations. They are mirroring the Devil within themselves. That is their god. Those within the veil have faith and practice from inside. A GOOD GOD who doesn't kill is their mirror.

(Gnostic Gospel of Philip) "None can see himself either in water or in a mirror without light. Nor again can you see in light without mirror or water."

CODEX II Translated by Wesley W. Isenberg Selection made from James M. Robinson, ed., The Nag Hammadi Library, revised edition. HarperCollins, San Francisco, 1990.

GOD didn't give the people a veiled Bible and cause them to err. They were already behaving those ways. The LORD designed the Bible to match their behavior. They accepted it. The WORD is then unveiled in their presence. Their own hearts translated their meaning.

(1 Corinthians 3:13) "each one's work will become clear; for the Day will declare it, because it will be revealed by fire; and the fire will test each one's work, of what sort it is."

(Revelation 2:23) "I will kill her children with death, and all the churches shall know that I am He who **search**es the minds and hearts. And I will give to each one of you according to your works."

Bowl of the Simeonite Church of Sardis

The Bible is a set of covenants. The first covenant is about the LORD telling the people to draw near to HIM. One thing HE advised was not to eat pork. Pork isn't healthy to eat. Scientific research later found that swine are about as smart as, or smarter than, a three-year-old child. Would these people eat their children? (Lamentations 2:20) That is three good reasons not to eat the animal. As if the LORD had his hand out, and the people were like wild animals, they refused to eat from HIS hand.

People wouldn't accept the truth. The Word of GOD had to enter within a veil.

The Old Testament taught obedience to GOD. The people were not allowed to just eat anything they wanted. They were also commanded not to join the carnal nations. It was said to have been an everlasting covenant, which the people refused. The story then shifted. Jesus was sent to save them since they didn't obey. The New Testament suggests that GOD would obey the people instead. It appeared that the LORD told the people to join the carnal nations and eat anything they wanted. They agreed and decided that it was from GOD. The LORD had sent HIS Son. HE allowed them to kill Jesus so that they could be in charge instead. That covenant was also said to have been everlasting. Jesus is the Word of GOD. They killed the truth and veiled it. He later resurrects through the unveiling.

The New Testament was basically about people living the same way that they already were. They agreed to live the new covenant because it meant that they didn't really have to listen to GOD. (Isaiah 9:17) Contrary to GOD's original plan, governments were quickly given charge over the churches. That's why they said that Satan ruled the Earth. (1 John 15:19)

The LORD used a veil on people because they wouldn't listen to HIM. GOD was within the veil. The Christians are the house of Israel. We know this because they say that the King of the Jews is their King. (Ezekiel 3:1-9) They were adopted. (Romans 8:15, 8:23, 9:4) (Galatians 4:5) (Ephesians 1:5)

The message is that the veiled Bible was the red stew that Jacob gave Edom. (Genesis 25:29-34) The stew (Word) was prepared and given to Edom in exchange for his birthright. This red stew offered meat within it. The birthright and blessing were then given to the vegans. The Christians were Edomites. (2 Esdras 1:33-37, 2:10-11) The red stew made Edom despise his birthright. The birthright was the astrology which they shunned. They can partake in the birthright if they believe in the Lamb. It is in Book 7 of this series, ready and prepared for the people.

An Old Testament law about diets was that linen was to be separated from wool. Linen wearers represents the vegans. Wool wearers (meat eaters) are not allowed to minister within the gates of GOD's inner court. (Ezekiel 44:17)

Bowl of the Judahite Church of Philadelphia

We know that the New Testament was veiled to the people so that the Bible could be delivered. If the Bible had directly said for the people to become vegan, the world would not have accepted it. Here is the evidence of this fact. These next quoted Gnostic verses explain that the womb of humanity would not accept Jesus unless He appeared as an animal like them.

(Gnostic Asclepius) "They found me, the son of the Majesty, in front of the womb which has many forms. I put on the beast, and laid before her a great request that heaven and earth might come into being, in order that the whole light might rise up. For in no other way could the power of the Spirit be saved from bondage except that I appear to her in animal form. Therefore she was gracious to me as if I were her son."

21-29 CODEX VI Translated by James Brashler, Peter A. Dirkse, and Douglas M. Parrott Selection made from James M. Robinson, ed., The Nag Hammadi Library, revised edition. HarperCollins, San Francisco, 1990.

The governments of the world are a dark womb. They wouldn't allow the birth of any spiritual child unless it mirrored their behavior. Jesus had to disguise Himself as a beast to get the Bible delivered. This allowed them to produce a spiritual child, yet that child would be idle. They wouldn't be able to translate the Bible. And they remained spiritually barren.

(Gnostic The Paraphrase of Shem) "It was separated from the light by the cloud of the silence. The cloud was disturbed. It was he who gave rest to the flame of fire. He humbled the dark womb in order that she might not reveal other seed from the darkness. He kept them back in the middle region of Nature in their position which was in the cloud. They were troubled since they did not know where they were. For still they (did> not possess the universal understanding of the Spirit. And when I prayed to the Majesty, toward the infinite Light, that the chaotic power of the Spirit might go to and fro, and the dark womb might be idle, and that my likeness might appear in the cloud of the Hymen, as if I were wrapped in the light of the Spirit which went before me, and by the will of the Majesty and through the prayer I came in the cloud in order that through my garment – which was from the power of the Spirit – the pleroma of the word, might bring power to the members who possessed it in the Darkness."

(Gnostic The Paraphrase of Shem) "For the soul is a work of unchastity and an (object of) scorn to the thought of Light. For I am the one who revealed concerning all that is unbegotten. And in order that the sin of Nature might be filled, I made the womb, which was disturbed, pleasant – the blind wisdom – that I might be able to bring (it) to naught."

The Paraphrase of Shem CODEX VII Translated by Frederik Wisse Selection made from James M. Robinson, ed., The Nag Hammadi Library, revised edition. HarperCollins, San Francisco, 1990.

(John 1:5) "And the light shines in the darkness, and the darkness did not comprehend it.

John the Baptist was the Archon of the Womb. When He baptized Jesus, the Word of GOD went under darkness and died. The Jordan River, which Jesus was baptized in, represented the desire for sexual intercourse with the dark womb. Society naturally desires to mate with the darkness. When the Ark of the LORD enters the Jordan, the will to sow seeds of darkness stops. (Joshua 3:13)

(Gnostic The Testimony of Truth) "The Jordan river is the power of the body, that is, the senses of pleasures. The water of the Jordan is the desire for sexual intercourse. John is the archon of the womb. And this is what the Son of Man reveals to us: It is fitting for you (pl.) to receive the word of truth, if one will receive it perfectly. But as for one who is in ignorance, it is difficult for him to diminish his works of darkness which he has done. Those who have known Imperishability, however, have been able to struggle against passions [...]."

The Hebrews had to go under the nations to be able to deliver the Seed of Abraham into the womb of society. John submerged Jesus into the Jordan. The Seed entered the people. Though the Seed was inside the dark womb, Jesus couldn't have a child with her. The dark womb was barren. She would deny those who could possibly be something good. Righteousness would deplete her ways. Like a black widow, she would also kill anyone that threatened her supreme authority. She wanted to be the only one. She (the carnal governments) thrived on war and honor. Jesus wasn't barren, yet the powers of the world were. The Hebrews had to hide the Oracle from them. Because of the veil, Jesus never made actual contact. The Bible was distributed. It went into a barren people. They were apostate. A light entered a darkness that didn't comprehend it.

(Gnostic The Tripartite Tractate) "The one who ran on high and the one who drew him to himself were not barren, but in bringing forth a fruit in the Pleroma, they upset those who were in the defect. Like the Pleromas are the things which came into being from the arrogant thought, which are their (the Pleromas') likenesses, copies, shadows, and phantasms, lacking reason and the light, these which belong to the vain thought, since they are not products of anything."

(Gnostic The Tripartite Tractate) "They were brought to a lust for power in each one of them, according to the greatness of the name of which each is a shadow, each one imagining that it is superior to his fellows. The thought of these others was not barren, but just like <those> of which they are shadows, all that they thought about they have as potential sons; those of whom they thought they had as offspring. Therefore, it happened that many offspring came forth from them, as fighters, as warriors, as troublemakers, as apostates."

To return Jesus from the depths of baptism, the Bible gets unveiled. His lights turn back on. All the religions from GOD become saved.

(Gnostic The Tripartite Tractate) "The prayer of the agreement was a help for him in his own return and (in that of) the Totality, for a cause of his remembering those who have existed from the first was his being remembered. This is the thought which calls out from afar, bringing him back. All his prayer and remembering were numerous powers according to that limit. For there is nothing barren in his thought. The powers were good and were greater than those of the likeness. For those belonging to the likeness also belong to a nature of falsehood."

CODEX I Translated by Harold W. Attridge and Dieter Mueller Selection made from James M. Robinson, ed., The Nag Hammadi Library, revised edition. HarperCollins, San Francisco, 1990.

The governments were the womb of darkness. They didn't want light; they wanted authority, war, and honor. That is why Jesus appeared as a story that matched Cain's ways. The governments believe in the honor of death. To put the Bible within them, His story is that He would die as a sacrifice for everyone. War soldiers claim to be sacrificing their lives for the people. That mentality fuels the war. It is the way of Cain. Jesus appeared to be like them. This gravitated the darkness of the military mind towards Him.

To remain undefiled, the Hebrews designed the New Testament to seemingly suggest that it supported carnal governments and their laws. They never literally said that. They just allowed the people to decide what it meant. They used a mirror of light (life) and darkness (death). They spoke of two laws, not clarifying which one they meant. The Law of GOD is life. The carnal laws of the nations are death. It was a test to see what they would choose.

(Deuteronomy 30:19) "I call heaven and earth as witnesses today against you, *that* I have set before you life and death, blessing and cursing; therefore choose life, that both you and your descendants may live;"

When Jesus died, all those who joined the governments died with Him. (Romans 6:5-10) (2 Timothy 2:11-13) They will also resurrect when the Word of GOD comes back to life within them. (1 Thessalonians 4:14) This barren dark womb is called Sophia. Those who entered her have been barren to the light.

(Gnostic The Gospel of Philip) "They called Sophia "salt". Without it, no offering is acceptable. But Sophia is barren, without child. For this reason, she is called "a trace of salt".

CODEX II Translated by Wesley W. Isenberg Selection made from James M. Robinson, ed., The Nag Hammadi Library, revised edition. HarperCollins, San Francisco, 1990.

We find that Sophia, like a black widow, wanted to produce something without her man. To be with the man means to be with the LORD our GOD. (Isaiah 54:5) (Jeremiah 3:20, 31:32) (Hosea 2:16) For this reason, the New Testament was designed to hide Jesus' literal dad. Being without a dad was appealing. It had reminiscence of when Eve was guiding the way. That's what the people liked. What they received was a doctrine of the dust (matter) of the earth. We see in this next verse that the kingdom that the Christians received was androgynous. That means that it was partly with GOD and partly without GOD. They had placed a doctrine from GOD into a government that wasn't from GOD. (2 Samuel 18:9)

(Gnostic The Hypostasis of the Archons) "Sophia, who is called Pistis, wanted to create something, alone without her consort; and her product was a celestial thing. A veil exists between the world above and the realms that are below; and shadow came into being beneath the veil; and that shadow became matter; and that shadow was projected apart. And what she had created became a product in the matter, like an aborted fetus. And it assumed a plastic form molded out of shadow, and became an arrogant beast resembling a lion. It was androgynous, as I have already said, because it was from matter that it derived."

The Reality of the Rulers CODEX II Translated by Bentley Layton Selection made from James M. Robinson, ed., The Nag Hammadi Library, revised edition. HarperCollins, San Francisco, 1990.

Only when the dad is revealed does it become a real spiritual matter from above the veil. Jesus biological dad isn't GOD, yet he is a seal of the unity of truth. His real dad is revealed in Book 7 of this series.

There is a womb of darkness. There is also a womb of light. Pistis Sophia is a mirror having both. Depending on which womb of hers that you choose, you may plant your seed. Her dark womb is outside the veil, barren to the light. Her light womb is within the veil, barren to the darkness. When planting the true Seed of Abraham, it can only be planted into the light womb.

(Gnostic The Thunder, Perfect Mind. I was sent forth from the power, and I have come to those who reflect upon me, and I have been found among those who seek after me. Look upon me, you who reflect upon me, and you hearers, hear me. You who are waiting for me, take me to yourselves. And do not banish me from your sight. And do not make your voice hate me, nor your hearing. Do not be ignorant of me anywhere or any time. Be on your guard! Do not be ignorant of me. For I am the first and the last. I am the honored one and the scorned one. I am the whore and the holy one. I am the wife and the virgin. I am <the mother> and the daughter. I am the members of my mother. I am the barren one and many are her sons. I am she whose wedding is great, and I have not taken a husband. I am the midwife and she who does not bear. I am the solace of my labor pains. I am the bride and the bridegroom, and it is my husband who begot me.

CODEX VI Translated by George W. MacRae Selection made from James M. Robinson, ed., The Nag Hammadi Library, revised edition. HarperCollins, San Francisco, 1990.

John's baptism represents the entering of the dark womb. He was later beheaded. That is why Christians taught like a mental lobotomy. Christians are trained not to think in any other direction. Their path becomes the only way that they can consider. They shut down. The testimony of John is that entering the dark womb beheads the people. That means that they become apostate. Entering the dark womb is to eat from the forbidden fruit.

(Gnostic The Testimony of Truth) "John was begotten by the World through a woman, Elizabeth; and Christ was begotten by the world through a virgin, Mary. What is (the meaning of) this mystery? John was begotten by means of a womb worn with age, but Christ passed through a virgin's womb. When she had conceived, she gave birth to the Savior. Furthermore, she was found to be a virgin again. Why, then do you (pl.) err and not seek after these mysteries, which were prefigured for our sake? It is written in the Law concerning this, when God gave a command to Adam, "From every tree you may eat, but from the tree which is in the midst of Paradise do not eat, for on the day that you eat from it, you will surely die."

CODEX IX Translated by Søren Giversen and Birger A. Pearson Selection made from James M. Robinson, ed., The Nag Hammadi Library, revised edition. HarperCollins, San Francisco, 1990.

Mary Magdalene is the Light Womb. She was a Benjamite. The name Benjamin means son of the right hand. Benjamites are known for their unwillingness to mate with the governments. (Matthew 25:33-41) Those who chose the dark womb are blind to the light. They are the left hand. We see within this next quote that Jesus chose to have children of light. He chose Mary because she was peaceful and light-loving.

(Gnostic The Gospel of Philip) "As for the Wisdom who is called "the barren," she is the mother of the angels. And the companion of the [...] Mary Magdalene. [...] loved her more than all the disciples, and used to kiss her often on her mouth. The rest of the disciples [...]. They said to him "Why do you love her more than all of us?" The Savior answered and said to them, "Why do I not love you like her? When a blind man and one who sees are both together in darkness, they are no different from one another. When the light comes, then he who sees will see the light, and he who is blind will remain in darkness." The Lord said, "Blessed is he who is before he came into being. For he who is, has been and shall be."

CODEX II Translated by Wesley W. Isenberg Selection made from James M. Robinson, ed., The Nag Hammadi Library, revised edition. HarperCollins, San Francisco, 1990.

Mary Magdalene was called male. To be a man means to be with GOD. The use of the words man and woman differ to mean "Did you follow Adam's first command?" or "Did you follow Eve?" Comprehend the way that gender meanings are being used. Sometimes they only mean gender.

(Gnostic Gospel of Thomas 114:2-3) "Jesus said: "Look, I will draw her in so as to make her male, so that she too may become a living male spirit, similar to you." (3) (But I say to you): "Every woman who makes herself male will enter the kingdom of heaven."

CODEX II Translated by Stephen Patterson and Marvin Meyer Selection from Robert J. Miller, ed., The Complete Gospels: Annotated Scholars Version. (Polebridge Press, 1992, 1994).

Those who sewed their seeds into the dark womb (female womb) had intercourse unto perdition. Their winds (voices) became supportive of the honor of war. Their path is sterile and barren.

(Gnostic The Paraphrase of Shem) "But the winds, which are demons from water and fire and darkness and light, had intercourse unto perdition. And through this intercourse the winds received in their womb foam from the penis of the demons. They conceived a power in their womb. From the breathing the wombs of the winds girded each other until the times of the birth came. They went down to the water. And the power was delivered, through the breathing which moves the birth, in the midst of the practice. And every form of the birth received shape in it. When the times of the birth were near, all the winds were gathered from the water which is near the earth. They gave birth to all kinds of unchastity. And the place where the wind alone went was permeated with the unchastity. Barren wives came from it and sterile husbands. For just as they are born, so they bear.

The Paraphrase of Shem CODEX VII Translated by Frederik Wisse Selection made from James M. Robinson, ed., The Nag Hammadi Library, revised edition. HarperCollins, San Francisco, 1990.

The Ancient Rigveda Hymns reveal the Abrahamic families of Shem, Ham, and Japheth. There are three wheels and a triple seat. They can only be correctly sealed into the light womb. The dark womb is called the she-wolf. When the kingdom was sealed into the she-wolf, which is the governments, the religions fought with each other for power. Details about this are revealed in Book 8 of this series. Here is a Rigveda hymn mentioning the kingdom and the two wombs.

([01-183] Hymn CLXXXIII Asvins 1-2) MAKE ready that which passes thought in swiftness, that hath three wheels and triple seat, ye Mighty, Whereon ye seek the dwelling of the pious, whereon, threefold, ye fly like birds with pinions. Light rolls your easy chariot faring earthward, what time, for food, ye, full of wisdom, mount it. May this song, wondrous fair, attend your glory: ye, as ye travel, wait on Dawn Heaven's Daughter.

([01-183] Hymn CLXXXIII Asvins 4) Let not the wolf, let not the she-wolf harm you. Forsake me not, nor pass me by or others. Here stands your share, here is your hymn, ye Mighty: yours are these vessels, full of pleasant juices.

The Hymns of the Rigveda – Translated by Ralph T. H. Griffith – 2md Edition, Kotagiri (Nilgiri) 1896

This doctrine is the Seed of the Light Womb (male womb). Here is the good mirror of Sophia. This is the light womb of doctrine.

(Gnostic Trimorphic Protennoia) "I am the Image of the Invisible Spirit, and it is through me that the All took shape, and (I am) the Mother (as well as) the Light which she appointed as Virgin, she who is called 'Meirothea', the incomprehensible Womb, the unrestrainable and immeasurable Voice.

CODEX XIII Translated by John D. Turner Selection made from James M. Robinson, ed., The Nag Hammadi Library, revised edition. HarperCollins, San Francisco, 1990.

Every action is either sewn into a womb of darkness or light. This isn't called the desire for sexual intercourse. Doctrinal desire for sexual intercourse means to remove virginity by mating with the carnal governments. Doing that would be partaking from the forbidden fruit.

(Gnostic Testimony of Truth) "When she had conceived, she gave birth to the Savior. Furthermore, she was found to be a virgin again. Why, then do you (pl.) err and not seek after these mysteries, which were prefigured for our sake? It is written in the Law concerning this, when God gave a command to Adam, "From every tree you may eat, but from the tree which is in the midst of Paradise do not eat, for on the day that you eat from it, you will surely die."

CODEX IX Translated by Søren Giversen and Birger A. Pearson Selection made from James M. Robinson, ed., The Nag Hammadi Library, revised edition. HarperCollins, San Francisco, 1990.

The womb of darkness had four breasts of Sodom and Gomorrah. These are the four groups of doctrine from the four riverheads that came from Eden. When they were planted into the dark womb, they became as children of Egypt. When the Bible is unveiled, all four breasts are replanted into the light womb.

(Gnostic The Gospel of the Egyptians) "Then there came forth from that place the great power of the great light Plesithea, the mother of the angels, the mother of the lights, the glorious mother, the virgin with the four breasts, bringing the fruit from Gomorrah, as spring, and Sodom, which is the fruit of the spring of Gomorrah which is in her. She came forth through the great Seth. Then the great Seth rejoiced about the gift which was granted him by the incorruptible child. He took his seed from her with the four breasts, the virgin,"

Fifth Commandment: Honor your father and mother. To honor your parents, you would need to walk the correct path. If your parents were incorrectly guided, and you follow them, that dishonors them. They may not know that. Honor your parents by doing the correct thing even if they weren't. Remember that they are not to be shamed. It isn't shameful to have been where and what they were. This is a no-shame doctrine. We face the facts and make the changes. We don't live in shame, nor should they. They got us here. There is another way to look at this. Jesus said not to call anyone on earth your Father. This means that we are to honor GOD.

Fourth Commandment: Keep the Sabbath day holy. When we have Sabbath meetings, we are not to have carnal officers educating the children. We are walking GOD's path. Only after someone has accepted these facts should they be leading the meetings. Someone who was once an officer can be in charge of a church. The person must first face the fact that he or she wasn't on the correct path. They can drop all shame. None are to shame them. They are forgiven. They must then pass through the training realignment in the Unveiled Living Law of Moses. After completing the requirements, they can educate in a religion. They become fully cleansed and exonerated. That's how turn their darkness into light. The other way separates the natural and spiritual realms by annulling the Spirit.

Bowl of the Naphtalite Church of Laodicea

Start at the foundation. In the beginning, Satan, the serpent, told Eve to eat the forbidden fruit. Serpents climb trees. They eat mice and rats. Eve allowed the serpent, which only eats meat, to tell her what to eat. It became her instructor. People began eating meat, and they became like beasts.

Then look forward in time to the New Testament days. Why do the Christians claim that meat-eating is GOD's way? Consider this. Jesus called Peter Satan. (Matthew 16:23) Peter then told everyone that he had a vision, whereupon he was told that he could eat anything. The vision had a sheet bringing animals for him to kill. (Acts 10:9-16) The sheet would have been linen, which signifies vegans. This sheet symbolizes Jacob (the vegans) handing the red stew for Edom to Peter. Then the church was said to have been founded on Peter, who is thought of as a rock. (Matthew 16:18) This rock is the rock by which Jesus would stone the churches and their leaders as false prophets. They wouldn't accept the Jews or Jesus any other way.

Many churches say that since the day of Paul, there hasn't been a single prophet. This happened because Paul helped Peter teach the people to eat anything they wanted. That path shuts GOD out.

The Word of GOD is said to be alive. This means that the Bible is doing things like a living being. (Hebrews 4:12) The Epistles of Paul were veiled. They are alive and well. The story is that Paul killed Christians until he was converted. Paul's change hasn't happened in the churches yet as of 2024. To convert Paul, the Bible must be unveiled. Once the Bible is unveiled, the Epistles of Paul don't kill people anymore. The Epistles were killing Christians by guiding people to eat anything they wanted. Peter and Paul gave the people an excuse not to listen to the truth. Do you see that the Bible is living?

Bowl of the Asherite Church of Galatia

Paul was blind for a time. He didn't eat or drink for three days. (Acts 9:1-10) Those three days were the 3,000 years between the times of Solomon and now. (2 Peter 3:8) This means that the Word of GOD was crucified during the days of Solomon when the ten tribes joined the nations. The Bible was then delivered to the people 1,000 years later. It was distributed as Jesus, the Word, likened to death. Let it be resurrected.

Jesus is the root and offspring of David. Jesus is the Word and the lamb. The lamb signifies vegan.

(Revelation 22:16) ""I, Jesus, have sent My angel to testify to you these things in the churches. I am the Root and the Offspring of David, the Bright and Morning Star.""

Here is one of the veils of King David.

David went vegan. That is what it meant by David having taken that one lamb. The lamb was said to have been Bathsheba, who was married to Uriah. (2 Samuel 12:1-15) David was Uriah, and Uriah was David's old self. He put himself on the frontline, killed the old man that he was, and renewed himself in the Mind of Christ. David took the light of the nation because the nation had the friction of war as their light. That is a natural, materialistic light. It was a murderer's light. (Job 10:22) By becoming vegan, David gained the Ark of the LORD GOD. (1 Kings 2:26) We can see this fact in Psalm 51.

Psalm 51 is titled: "To the Chief Musician. A Psalm of David when Nathan the prophet went to him, after he had gone in to Bathsheba."

(Psalm 51:4) "Against You, You only, have I sinned, And done *this* evil in Your sight—That You may be found just when You speak, *And* blameless when You judge.

What we find is that when David killed Uriah, he didn't claim wrong against Uriah; he claimed wrong only against the LORD. If David had killed someone for their wife, then he would have written that he had sinned against that man. He asked the LORD to cleanse him from the guilt of bloodshed. His mouth could then praise GOD correctly.

(Psalm 51:4) "Deliver me from the guilt of bloodshed, O God, The God of my salvation, *And* my tongue shall sing aloud of Your righteousness. O Lord, open my lips, And my mouth shall show forth Your praise."

David was saying that he was conceived as a meat eater. The LORD brought David into veganism.

(Psalm 51:5) "Behold, I was brought forth in iniquity, And in sin my mother conceived me."

The ewe lamb that David took was Bathsheba. When she went vegan, her ability to connect to her family greatly departed. (2 Samuel 12:1-3) This was hard for her and her family to bear. They were not vegan, and she was no longer with them in that area of life. The wrong was put away because David didn't actually do anything wrong.

Veganism was like a sword against David's family. The people didn't want that path. They hated it. (Psalm 45:1-15) It became a time of peace. They were not to kill animals anymore. Those like Joab refused veganism. His belt was made of animal skin. His feet remained on a crooked path.

(1 Kings 2:5) ""Moreover you know also what Joab the son of Zeruiah did to me, *and* what he did to the two commanders of the armies of Israel, to Abner the son of Ner and Amasa the son of Jether, whom he killed. And he shed the blood of war in peacetime, and put the blood of war on his belt that *was* around his waist, and on his sandals that *were* on his feet.

While the people who refused were in darkness, the LORD accepted David's sacrifice. His sacrifice matched the offering of Abel. The lamb offered by Abel meant that he gave up eating meat for the LORD. The plant offering by Cain meant that he gave up eating fruits and vegetables for the LORD.

David, because of his obedience, gained the rights to the throne of the Ark. In a way he seemingly did take it from the other tribes. He didn't do it on purpose. It really was GOD's plan.

There are two main things that distinguish a Canaanite: meat eating and working for the carnal government. As scientifically proven in chapter four of this book, Jerusalem means Earth. Jerusalem was also considered the land of Canaan. In this aspect, David was a Hittite of Canaan.

(Ezekiel 16:3) "Again the word of the LORD came to me, saying, "Son of man, cause Jerusalem to know her abominations, and say, 'Thus says the Lord GOD to Jerusalem: "Your birth and your nativity *are* from the land of Canaan; your father *was* an Amorite and your mother a Hittite."

There is another veil to David's actions. In this one, Uriah really is another man. Uriah was a real Canaanite who worked for the government. His image was that he was David's best man. That image was false. The story begins like this. Joab was the basic veiled Bible. Those who used the Bible to justify joining the military were with Joab. Bathsheba was a prisoner of war. The wives of men who join the military, who grab their guns, and enter another man's property are prisoners of war. This is the Law of GOD. (Deuteronomy 21:10-14) Uriah had entered the land of another man and had threatened him. He was part of the Canaanite army. They were against GOD. Christians today teach to stand up for their nation, which encourages joining the military. Theirs is the path of the Joab. Those in the military are walking the path of Uriah.

(2 Samuel 11:6) "Then David sent to Joab, *saying,* "Send me Uriah the Hittite." And Joab sent Uriah to David."

When Uriah came to David (the Bible), the king told him to go home and wash his feet. This means to put his guns down. He was walking the crooked path.

(2 Samuel 11:8-10) "And David said to Uriah, "Go down to your house and wash your feet." So Uriah departed from the king's house, and a gift *of food* from the king followed him. But Uriah slept at the door of the king's house with all the servants of his lord, and did not go down to his house. So when they told David, saying, "Uriah did not go down to his house," David said to Uriah, "Did you not come from a journey? Why did you not go down to your house?"

Uriah wouldn't put his gun down and go home. Since his wife was a prisoner of war, she was legally David's. GOD gave Bathsheba to King David. We see in this next verse that the Ark (David) and real Israel and Judah were in (with) their tents (wives). The people of Joab and Uriah were encamped in the lands of another.

(2 Samuel 11:11) "And Uriah said to David, "The ark and Israel and Judah are dwelling in tents, and my lord Joab and the servants of my lord are encamped in the open fields. Shall I then go to my house to eat and drink, and to lie with my wife? *As* you live, and *as* your soul lives, I will not do this thing."

Since Uriah wouldn't listen, David let him remain in his ways. Uriah was drunk in the veiled ways of the Bible. (Ephesians 5:18) He would then lie in the bed of the churches, helping teach that the military and the nation were GOD's path. Uriah would have done that either way. He still wouldn't put his gun down.

(2 Samuel 11:12-13) "Then David said to Uriah, "Wait here today also, and tomorrow I will let you depart." So Uriah remained in Jerusalem that day and the next. Now when David called him, he ate and drank before him; and he made him drunk. And at evening he went out to lie on his bed with the servants of his lord, but he did not go down to his house.

Uriah (the carnal officers) refused to put their weapons away. Because of this, David (the Bible) put Uriah on the front lines of the hottest battle. This battle was like the ones that North America fought against the Muslims between 1990 and 2019. Throughout that war, the North American military was on the Muslim's land, threatening them. During those days, the Christians, like Joab, kept the military (Uriah) on the frontlines. The Christians joyfully fed the battle while the Muslims were a decoy antichrist.

(2 Samuel 11:14-17) "In the morning it happened that David wrote a letter to Joab and sent *it* by the hand of Uriah. And he wrote in the letter, saying, "Set Uriah in the forefront of the hottest battle, and retreat from him, that he may be struck down and die." So it was, while Joab besieged the city, that he assigned Uriah to a place where he knew there *were* valiant men. Then the men of the city came out and fought with Joab. And *some* of the people of the servants of David fell; and Uriah the Hittite died also."

Tenth Commandment: You shall not covet your neighbor's house, wife, or property. North America was hiding the fact that they were coveting the Middle Eastern oil. North America uses more than anyone else. The Middle East had the most oil in that day. When a soldier enters another man's land like that, his wife is not considered his wife. She is considered by the Law of GOD, a prisoner of war. They were on Muslim land. This Law from GOD is explained in the Qur'an and the Bible. (Women Tested Surah 60:7-11)

(2 Samuel 11:25) "Then David said to the messenger, "Thus you shall say to Joab: 'Do not let this thing displease you, for the sword devours one as well as another. Strengthen your attack against the city, and overthrow it.' So encourage him."

(Revelation 3:10) "He who leads into captivity shall go into captivity; he who kills with the sword must be killed with the sword. Here is the patience and the faith of the saints."

At that point, David had displeased the LORD. Because of this, his first child had to die. David's first child was the veiled Bible. That path of the military mind displeases the LORD. That is the path that the veiled Bible teaches. David had to change.

In a spiritual sense, David was one with Uriah (Canaan). Everyone usually was at birth. Once David transformed, he entered the veil and became a new man. He changed into a vegan and became the Ark of GOD. The Ark would pass from generation to generation in his lineage. Each person who was called would get very close to the LORD. David's sons would work for GOD. The good sword (sharpened Word of GOD) would be with David's lineage always. David had done this secretly with a veil. He had done some something good, not bad. The veil was only bad because the people were that way already.

(2 Samuel 12:1012) "Now therefore, the sword shall never depart from your house, because you have despised Me, and have taken the wife of Uriah the Hittite to be your wife.' Thus says the LORD: 'Behold, I will raise up adversity against you from your own house; and I will take your wives before your eyes and give *them* to your neighbor, and he shall lie with your wives in the sight of this sun. For you did *it* secretly, but I will do this thing before all Israel, before the sun.' ""

When a man is that close to GOD, women often have a very hard time being with them. There are women who will accept a man when he works for the LORD. That is very hard to find, though. Solomon went through the same thing. He sought a woman who would be true to GOD his entire life. The females wanted something else. Here are Solomon's words.

(Ecclesiastes 7:27-28) ""Here is what I have found," says the Preacher, "*Adding* one thing to the other to find out the reason, Which my soul still seeks but I cannot find: One man among a thousand I have found, But a woman among all these I have not found."

Solomon was also without sin. His veiled doctrine had married all the governments. He didn't actually do that. He wasn't a bad person. Under the veil, the LORD was happy with him. The throne has been restored.

David often thought that he had done something wrong. He didn't do anything bad. Only an enemy of the LORD would blaspheme him. Bathsheba was given to David by GOD. She was drawn to him. By Law, she had every right to make the decision not to be a prisoner of war.

(2 Samuel 12:13-14) "So David said to Nathan, "I have sinned against the LORD." And Nathan said to David, "The LORD also has put away your sin; you shall not die. However, because by this deed you have given great occasion to the enemies of the LORD to blaspheme, the child also *who is* born to you shall surely die.""

There isn't anyone who can get into the doctrinal house of David without being vegan. Meat eaters and the military mind have attempted to enter the back door every time.

(John 10:102) ""Most assuredly, I say to you, he who does not enter the sheepfold by the door, but climbs up some other way, the same is a thief and a robber. But he who enters by the door is the shepherd of the sheep.

Seventh Commandment: You shall not commit adultery. The church is called Jesus Bride. Those who join the governing forces of the carnal nations are considered adulterers. They are unfaithful to GOD. They are those who kill the prophets' truth. (Revelation 17:1-6) And the nations are not to covet the religions that are Jesus' bride.

(Gnostic The Exegesis on the Soul) "For from them she gained nothing except the defilements they gave her while they had sexual intercourse with her. And her offspring by the adulterers are dumb, blind and sickly. They are feebleminded."

(Gnostic The Exegesis on the Soul) "Then the bridegroom came down to the bride. She gave up her former prostitution and cleansed herself of the pollutions of the adulterers, and she was renewed so as to be a bride. She cleansed herself in the bridal chamber; she filled it with perfume; she sat in it waiting for the true bridegroom."

(Gnostic The Exegesis on the Soul) ""Hear, my daughter, and see and incline your ear and forget your people and your father's house, for the king has desired your beauty, for he is your lord." For he requires her to turn her face from her people and the multitude of her adulterers, in whose midst she once was, to devote herself only to her king, her real lord, and to forget the house of the earthly father, with whom things went badly for her, but to remember her father who is in heaven."

CODEX II Translated by William C. Robinson Jr. Selection made from James M. Robinson, ed., The Nag Hammadi Library, revised edition. HarperCollins, San Francisco, 1990.

Bowl of the Issacharite Church of Philemo

The Bible states that the people draw near to GOD with their mouth and honor with their lips. These people honor with their words, yet are not really with GOD. They claim GOD while they speak falsely and walk spiritually backwards. (Isaiah 29:13) (Jeremiah 7:24)

GOD revealed that people will eat of every moving thing that lives and that they will pay a blood price for every animal that they eat.

(Genesis 9:3-5) "Every moving thing that lives shall be food for you. I have given you all things, even as the green herbs. But you shall not eat flesh with its life, *that is,* its blood. Surely for your lifeblood I will demand *a reckoning;* from the hand of every beast I will require it, and from the hand of man. From the hand of every man's brother I will require the life of man.

The veiled story is that the Christians offered their king in exchange for the rights to eat meat. They wanted meat so much that they traded their king for the blood price.

When GOD said that every moving thing that lives shall be food for them, humanity was also moving as the trees. This means that people would destroy each other. Their destruction is caused by carnal forces being deployed in the people. War is a blood price for eating meat. Those armed forces block GOD.

(Isaiah 7:2) "And it was told to the house of David, saying, "Syria's forces are deployed in Ephraim." So his heart and the heart of his people were moved as the trees of the woods are moved with the wind."

The blood price paid for eating meat became a mirror. The reflection is that people kill people. (1 Corinthians 13:12) (2 Corinthians 3:18) Their offerings become a sickness as well as their incense (prayers). Compare these verses and see the blood price for eating meat.

(Genesis 9:6-7) ""Whoever sheds man's blood, By man his blood shall be shed; For in the image of God He made man. And as for you, be fruitful and multiply; Bring forth abundantly in the earth And multiply in it.""

(Isaiah 66:3-4) ""He who kills a bull *is as if* he slays a man; He who sacrifices a lamb, *as if* he breaks a dog's neck; He who offers a grain offering, *as if he offers* swine's blood; He who burns incense, *as if* he blesses an idol. Just as they have chosen their own ways, And their soul delights in their abominations, So will I choose their delusions, And bring their fears on them; Because, when I called, no one answered, When I spoke they did not hear; But they did evil before My eyes, And chose *that* in which I do not delight.""

While Moses was on Mount Sinai getting the commandments from GOD, the people made a golden calf. That story is universal. (Isaiah 42:9) This calf was the people's interpretations of the Word of GOD. Their views of Jesus lead to harmful laws and wars. They had taken in a veiled doctrine. We find that the churches received the golden calf. Since the Word of GOD is Jesus, the churches continuously kill Him every day. (1 Corinthians 15:31) They would have had it no other way.

(Barnabas 8:2) "Understand ye how in all plainness it is spoken unto you; the calf is Jesus, the men that offer it, being sinners, are they that offered Him for the slaughter. After this it is no more men (who offer); the glory is no more for sinners."

Bowl of the Zebulunite Church of Thessalonia

By learning to kill animals, humans were cultured to kill people. Jesus was vegan. He was the Lamb which is a symbol of veganism. He, the Word of GOD, was killed so that the people couldn't understand this. Humanity was murdering Him.

When approached by vegan advice, Christians would often hide behind verses stating that Jesus ate fish. (Luke 24:42-43) Jesus fed over 5,000 people with five loaves of bread and two fish. How did He do that? (Matthew 14:17-21)

The Bible is a written cipher, and Jesus was a fisher of men. The people are his fish. (Matthew 4:19) Being that the fish are people, to eat the fish means to read them and/or their writings. When Jesus was resurrected, he ate fish. This means that He read the Word of GOD. He is the teacher, and His fish swim in schools.

Jesus broke bread to feed thousands of people. The bread was the word of GOD. He shared the Word with the masses, and they were all filled with life. He wasn't speaking about bread.

(Matthew 16:9-12) "Do you not yet understand, or remember the five loaves of the five thousand and how many baskets you took up? Nor the seven loaves of the four thousand and how many large baskets you took up? How is it you do not understand that I did not speak to you concerning bread?—*but* to beware of the leaven of the Pharisees and Sadducees." Then they understood that He did not tell *them* to beware of the leaven of bread, but of the doctrine of the Pharisees and Sadducees."

(John 4:34) "Jesus said to them, "My food is to do the will of Him who sent Me, and to finish His work.""

(John 6:48-51) "I am the bread of life. Your fathers ate the manna in the wilderness, and are dead. This is the bread which comes down from heaven, that one may eat of it and not die. I am the living bread which came down from heaven. If anyone eats of this bread, he will live forever; and the bread that I shall give is My flesh, which I shall give for the life of the world.""

Jesus' flesh means unveiled doctrine. (1 John 4:3)

Jesus can break the bread of life to feed as many people as are present. Because Jesus is the bread of life and the Word of GOD. (John 1:1-18) His flesh sustains the life of people and the Earth.

When Jesus feeds (reads) fish (flesh) to the people, it means that He is reading (feeding) them the unveiled Word of GOD.

Jesus was the bread from heaven. The people ate his interpretations. When He read and properly interpreted the books of the Bible, people were filled. They were satisfied with his interpretations. (Matthew 13:10-15) (Proverbs 1:6) Those who weren't His disciples were most likely not satisfied.

Bowl of the Gadite Church of Corinth

Would you like divine rights to heal? Do you pray to GOD asking to be healed? If you are killing animals to eat, then what do you deserve? You may consider your reasoning if you still eat meat. What is the righteous basis for consuming dead animals?

People harm themselves with a reflective outcome when they eat meat. Many doctrines, including Biblical and Hindu, teach that we are judged according to our ways. (Ezekiel 18:30, 24:14, 33:20) If GOD asked you why your body is sick and what it is that you eat, how would you reply? Do you provide the suffering and killing of many animals? Do you eat them because they taste good? GOD may answer you as in these verses. (Psalm 5:9, 51:6) (Luke 11:39) (Romans 3:13) (Philippians 3:18-19)

If you eat meat, then you are eating unhealthily for comfort while harming others. The internal workings of eating animals are known as selfishness. If you are not promoting the health of yourself and others, then is it right for you to ask GOD to heal you? (Zechariah 7:13)

Should GOD heal you to reverse the effects of that which you do to your surroundings? That contradicts the teachings of both GOD and Karma. One should first stop damaging their environments.

Jesus taught that nothing entering the body from the outside defiles it. (Matthew 15:10-20) If you eat meat, then out of you comes murder. Buying meat is like hiring a hitman to kill an animal and chop it up for you. The actions that come out of the heart to get the meat are murder and destruction. Also, eating meat causes the people to reason like an animal. Reasoning like animals leads to constant global warring. The meat industry destroys the world. Meat farms are devastating the atmosphere and the rivers. Here is a link to a documentary revealing many of the global problems from eating meat.

https://youtu.be/YbfXtcaJ7AU

There is also important vegan information in the video in that link.

Every time you eat meat, a bad example comes out of you, which perverts the ways of the people. Now consider the difference of when you put veganism into you. Being vegan helps heal the entire world, including the rivers and the atmosphere. People that think as a vegan often become far more peaceful and efficient. People begin caring more about others and the animals. People become genuine and careful. The vegan standard is not to war with one another. Vegans will to protect the environment and the entire earth. Without murder entering, death doesn't abound.

If you are willing to listen and therefore produce yourself as a living and healing energetic and spiritual expression, then your prayers will reflect that worthiness to be heard. You will deserve to be answered. Listen, and you will be listened to. (Proverbs 27:19) (Psalm 66:18-20) (Isaiah 66:3-4)

Begin by not hiding from the LORD anymore. The LORD will no longer hide from you. To do this, enter veganism. Enter the veil. (Ezekiel 39:29) (Revelation 6:15-16)

There are both spiritual and physical benefits to being vegan. When we are vegan, we no longer have to pay the blood price of killing the animals. GOD will heal us.

For the churches to satisfy GOD, they need to do two things to begin with.

1: They must become vegan.

Book 8 of this series has a vegan dietary plan. Some people may like to go further than a plant-based diet. That book has a Livet that can help. The Livet is a path for spiritual nutrition. It goes further than standard veganism, providing a no-kill system. Within the Livet, neither plants or animals are killed for food. When your intake is a no-kill system, your intake breathes full life.

2: They must not be associated with earthly governmental affairs of the military nations.

There is a message from Saint Peter about apostasy. To find it, we compare the Gospels. The Gospel of John is about 92% unique. The Gospel of Luke is about 59% unique. The Gospel of Matthew is about 42% unique. The Gospel of Mark is only about 7% unique. The reason that Mark had the least amount of revelation by far was because he was in Babylon as the churches have been. Entering the Babylonian ways of carnal governments leads to apostasy.

(1 Peter 5:13) "She who is in Babylon, elect together with *you,* greets you; and *so does* Mark my son."

The Bible tells us to come out of Babylon. Never fight them; merely disassociate from their ways.

(Revelation 18:4) "And I heard another voice from heaven saying, "Come out of her, my people, lest you share in her sins, and lest you receive of her plagues."

www.ingramcontent.com/pod-product-compliance
Lightning Source LLC
Chambersburg PA
CBHW061117170426
43199CB00026B/2950